Bragg Fibers

Ritesh Kumar Chourasia • Aavishkar Katti

Bragg Fibers

From Optical Properties to Applications

Springer

Ritesh Kumar Chourasia
Post-Graduate Department of Physics
Samastipur College (A Constituent
Unit of L.N.M.U. Darbhanga-846004)
Samastipur, Bihar, India

Aavishkar Katti
Department of Physics
Dr. Vishwanath Karad MIT World Peace
University
Pune, Maharashtra, India

ISBN 978-3-031-65163-2 ISBN 978-3-031-65164-9 (eBook)
https://doi.org/10.1007/978-3-031-65164-9

This Springer imprint is published by the registered company Springer Nature Switzerland AG
The registered company address is: Gewerbestrasse 11, 6330 Cham, Switzerland

I would like to dedicate this book to honor my parents (Mom and Dad), who planted the seed of knowledge, kindness, and human values in my mind and nurtured it. I bow my gratitude to my beloved wife and my heartbeat (lovable family members) for their continuous love and support. —Ritesh Kumar Chourasia

I would like to dedicate this book to Mom and Dad, who have been role models for me in my academic career, The Omniscient Lord Almighty, Lord Dattatreya, who is within and without. —Aavishkar Katti

Foreword

Recent Nobel prizes stand a testimony to the fact that optics and photonics are contributing greatly to the development of societies around the world. The invention of lasers in 1960 provided an unprecedented impetus to the field, and progress in this field has been very rapid leading to developments in fiber optics and optical communication, nonlinear optics, quantum optics, optical sensing, super-resolved imaging, and many other areas. The UN recognized these contributions with the declaration of 2015 as the Year of Light and Light-based Technologies, following with the declaration of May 16 as the International Day of Light. Academic and commercial activities, in this field, have also seen unprecedented growth over the last few decades. In particular, optical fibers, particularly specialty fibers, have provided a platform for many developments including sensors, nonlinear optics, and high power lasers. In this respect, fibers with some periodic variation in structures have become particularly interesting from the point of view of many applications. These include fiber gratings, photonic crystal fibers, and Bragg fibers. Some of these structures are inspired by structures in nature. This book is concerned with Bragg fibers which in the most basic form contain periodic or nearly periodic layers of high and low refractive indices in the radial direction. These have many unusual properties and require a somewhat involved mathematical framework to analyze. Since the initial proposal in late 1970s, a great amount of research and development work has been done in over four decades. This book is a commendable effort to describe and analyze these developments.

The book provides a comprehensive overview of the latest advancements in multi-layered, multi-material Bragg fibers spanning nine chapters. The first few chapters cover analytical techniques, theorems, and commonly used mathematical techniques. Next three chapters discuss various optical characteristics of these fibers and how these can be tailored. These are then followed by two chapters on applications of these fibers in sensors and other devices. The final chapter briefly includes some aspects of nonlinear wave propagation through Bragg fibers. The book thus presents a clear and coherent progression of the theoretical formulation, modeling, optical features, and practical advancements of multilayered, multi-material Bragg fiber. The authors have vast experience of research in the area and significant part of the book has grown out of their work. This gives the depth and clarity to the presentation.

This book, probably the first on this topic, would serve as a valuable reference and research monograph for both beginning and experienced researchers in the fields of theoretical and experimental optics and photonics with a particular interest in specialty fibers. This book would be very useful for graduates and postgraduate students in physical sciences who are interested in conducting research in the field of photonics.

This book is a welcome and valuable addition to the educational literature on the subject and would benefit students in science and engineering in learning about Bragg fibers and their applications.

IIT Delhi, New Delhi, Delhi, India Anurag Sharma

Preface

Total internal reflection, a well-known optical phenomena, is utilized to guide light in conventional optical fibers (COFs). In order to account for this optical phenomena, the core material of the COFs must possess a refractive index (RI) that is greater than that of the cladding material. The benefits of these COFs compared to traditional waveguides such as microwave devices and two-wire transmission lines are manifold: they provide a wider bandwidth with lower attenuation, exhibit resistance to electromagnetic (EM) interference, and possess characteristics of being lighter, more flexible, safer, and more secure. However, COFs do have disadvantages. The high refractive index (RI) of the core material leads to several unwanted consequences, including doubly degenerate fundamental modes, losses in the cladding and material, unidirectional propagation of light, strong nonlinearity, and geometrical and material dispersions. Therefore, these significant limitations of COFs can be overcome by innovative Bragg fibers (BFs) that have a hollow/air core. The publication of the influential study by Yeh and Yariv in 1976 ignited a surge of interest in the field of BFs research and continues to captivate researchers' focus to this day. The Bragg gratings (BFs), consisting of periodically arranged cylinders with varying refractive indices (RIs), play a crucial role in guiding the light through the waveguide by utilizing multiple Bragg reflections. The phenomenon of Bragg reflection enables the transmission of light through a hollow core. These devices have minimal nonlinearities, are not affected by polarization or birefringence, do not experience Fresnel reflections at the end of the fiber due to coupling in air, and are not influenced by material or cladding loss because over 98% of the mode is confined in the air rather than the high-index silica material. The hollow core of Bragg fiber facilitates the entrance or passage of biological solutions or chemical gases, rendering it an optimal platform for chemical or biological sensors. A solitary TE or TM mode can spread through BFs, and other means, due to the fact that the basic modes are not degenerate. Furthermore, the biological roots of the arrangement of cylindrical concentric periodic patterns in the shape are well established. The botanical fruit *Margaritaria nobilis* has an outward form resembling a cylindrical structure with layers. This form of construction is inspired by a multilayer artificial structure that possesses a photonic band gap (PBG), which may be further adjusted by applying mechanical pressure to its surface. The construction of a functional Photonic Band Gap (PBG) is significantly influenced by the periodic nature of the Bragg's Factor (BF) structure. The form, dimensions, arrangement, and coating material of the multilayer periodic structure together influence the effective width of the photonic bandgap (PBG). By employing photons as the means of transmitting information, instead of the less efficient electrons, the telecommunications sector will undergo a revolutionary transformation, resulting in a significant enhancement in the capacity and speed of intricate communication networks. However, in order to utilize the technological potential of BFs effectively, it is crucial to have the capacity to adjust and modify their characteristics dynamically. This book examines the transmission of electromagnetic waves in hollow-core Bragg fibers for the purpose of sensing and filtering. The proposed structure features a centrally hollow core, which is surrounded by cladding consisting of alternating layers with high and low refractive indices. The regular occurrence of alternate claddings with high and low refractive index in Bragg fibers results in the formation of a photonic bandgap. Researchers utilize this photonic bandgap (PBG) as a signal in sensing applications. In order to enhance the sensing capabilities, they employ cladding layers that possess a short photonic bandgap (PBG) and a low refractive index (RI) contrast. In order to improve the sensing capabilities, the structure of the BFs includes a defect layer that acts as a Fabry–Pérot resonator. When the defect is located in the appropriate photonic bandgap region, a prominent peak is noticed. This defect peak might potentially be utilized to create a narrow-band optical filter or a detection signal. An intelligent material placed within the cavity exhibits an electro-optic effect when subjected to an external electric field, resulting in the displacement of the defect peak within the analyzed photonic bandgap. The resonant state within the BFs structure fluctuates as a result of mechanical strain induced along the cavity. Therefore, the utilization of an external field allows for further adjustment of the suggested BFs structure. Furthermore, the proposed framework offers a wide range of potential applications in the future. These include the development of optoelectronic noses that utilize Boolean-olfaction and multi-material multifunctional BFs capable of visual perception, auditory reception, sensory detection, and communication. Additionally, the framework might enable the creation of square-shaped multi-material BFs, among other possibilities. The primary goal of this book is to ascertain the sensing applications of electromagnetic wave propagation characteristics in boundary fields with a defect layer. The book is organized into nine chapters.

Chapter 1 of this book explores the relationship between the physical characteristics, visual appearance, and chemical dependence of natural photonic systems. It demonstrates how this understanding may be applied to create new artificial photonic systems for various applications in photonics and sensing. Our strong motivation compels us to engage in natural photonics research, resulting in the development of specialized templates for photonic technology layouts. Here, we examine the findings of a study on the hierarchical optical structure present in the grain shell of *Margaritaria nobilis* fruits. Additionally, it resulted in the creation of pioneering Bragg reflected light BFs, which disturb a coaxially stacked structure within the individual cells of the grain's outer tissue layers. This bioinspired BFs structure has two technologically exploitable properties for modifying light and color. Wavelength-selective light scattering occurs due to the presence of both nanoscale regularity and microscale cylindrical symmetry, resulting in light being scattered in various directions. The artificial bio-inspired structure discussed in Chap. 1 is a multilayer-wrapped structure made up of two different refractive index phases and a hollow center. The artificial system simplifies certain complications seen in the natural structure, such as the elliptical shape of the fruit cell and the cross-section inside the periodic layers. Despite these simplifications, the artificial system maintains a similar scale and underlying optical interaction to its natural counterpart.

Chapter 2 explores the utilization of standard fiber drawing processes to construct BF structures from tiny preforms. However, the ability to draw materials into multilayer fibers is rather restricted. In order to reduce the occurrence of fractures at the interfaces of materials when operating at high temperatures, it is necessary for the pre-made material combinations to possess a suitable refractive index contrast and thermal expansion coefficients that fit the lattice structure. The chapter also explores an alternative approach to producing fibers at room temperature. This process involves using several types of organic polymers and inorganic materials, which provide a wide range of optical and mechanical qualities. Unlike thermally drawn fibers, these fibers are not restricted to having longitudinal symmetry along their axis. Additional dielectric materials utilized in fiber production encompass thermoplastics, such P.S. polystyrene and PMMA, which stands for polymethyl methacrylate. The rolling approach has been employed to incorporate thin metal films into non-stretchable fibers. Spray coating and blade coating are established industrial methods that may be used in roll-to-roll manufacturing and should be considered as potential possibilities for producing large-scale bilayers. During the process of film deposition, it is possible to modify the specific thicknesses of the two films in the first bilayer. Consequently, the position of the fibers' reflection band in the spectrum may be readily modified. These fibers display significant reflection over a limited range of wavelengths determined by the periodicity of the multilayer structure. This reflection is accompanied by a decrease in transmission, resembling the optical characteristics of the tropical fruit.

Chapter 3 employs the transfer matrix technique, the asymptotic method, and the Galerkin method to provide a theoretical description of the structure of BFs. In addition, it explores how the transfer matrix method may be modified to calculate reflection at interfaces, single slabs, and multilayer cylindrical stacks, as well as the dispersion optical characteristics of Bragg filters. This approach is more straightforward compared to the often employed Chew's method due to its reduced number of High/Low RI layers. It has been shown that in order to produce accurate results, the criterion of asymptotic approximation must be satisfied. The T.E. refers to the acronym for "Technology Engineering." The terms "H-Polarization" and "T.M." refer to certain modes of electromagnetic wave propagation. The study focuses on analyzing the E-Polarization modes and exploring the bandgap.

Symmetric birefringent fibers possess a core with a low refractive index that is hollow and surrounded by a cladding that is periodic and free from any geometric or symmetry imperfections. Chapter 4 of the book focuses on the characteristics of hollow-core BFs, including their propagation and dispersion properties. These BFs are designed to have high refractive index (RI) cladding contrast combinations and are optimized for minimal loss at the visible window wavelength of 632.8 nm. The boundary matching methodology has been employed to establish a correlation between the incident and transmitted light waves utilizing the most suitable transfer matrix technique. The photonic band gap (PBG) obtained in high contrast cladding Bragg fiber (HCCBF) has a significant magnitude, hence necessitating fewer periodic cladding layers to get a flawless PBG with complete reflection (100%). The width and position of photonic bandgaps (PBGs) are greatly influenced by the angle at which the electromagnetic wave strikes, specifically the route that the incoming light takes. In sensing applications, the sensitivity of the Bragg fiber may be assessed by measuring either the width or the spectrum shift of the PBG.

In Chap. 4, we observe that the high-contrast cladding hollow core BFs exhibit a wider photonic bandgap (PBG) with a significantly high full width at half maximum (FWHM). Thus, the performance of these setups is considerably affected in sensing applications. In Chap. 5, we recommend for achieving the smallest full width at half maximum (FWHM) of the sensor signal. To attain the lowest full width at half maximum (FWHM), one might utilize a defect layer or disrupt the symmetry of the Bragg periodic cladding. In this chapter, we construct and theoretically assess hollow core BFs with a defect layer for the design window at a wavelength of 632.8 nm. The connection between the initial and final fields has been stabilized by ensuring that the electric and magnetic fields are in alignment at different interfaces, as explained extensively in Chap. 3. The formulae for reflectance and transmittance have been discovered as well. A Bragg periodicity in the tested wavelength range results in the formation of a flawless photonic bandgap (PBG). The presence of a defect layer inside Bragg periodicity results in the emergence of a minor defect peak within the ideal photonic bandgap zone. The frequency at which the cladding layers repeat influences the full width at half maximum (FWHM) of this defect peak. The position and displacement of the defect peak are also influenced by the beam's angle of incidence and the refractive index of the core material. Consequently, rather than utilizing the full photonic bandgap (PBG) as a detecting signal, it is preferable to focus on the defect peak.

Chapter 6 focuses on the efficient design and manufacturing of BFs that are made from many materials and have multiple functions. The chapter specifically explores the use of preform-based technologies for this purpose. The fiber optic devices are composed of chalcogenide semiconducting glass and an insulating polymer bilayer. In addition, there are four metal electrodes available for the fabrication of novel integrated devices. Furthermore, it investigates the electrical and optoelectronic capabilities of metal-semiconductor-insulator BFs in order to fulfill the required device functionalities. The utilization of Bragg optical resonators with external surface reflectors on fiber resonators has also been investigated for hollow BFs core bases. Self-monitoring beamforming systems have the potential to be extremely important for transmitting high amounts of optical power. These systems may be used in various industries, such as industrial and medical sectors. Chapter 6 provides a comprehensive analysis of the application of narrowband Bragg photodetectors, which utilize amorphous chalcogenide glasses to accurately detect a wide spectrum of light wavelengths. This book chapter further investigates the coveted and unique use of surface-emitting lasers utilizing BFs.

Previously, we have observed that BFs possess a structure characterized by a hollow core, rendering them well-suited for applications in bio/chemo sensing. The hollow core of the BFs structure functions as a flow cell, capable of containing biological solutions or facilitating the movement of chemical gasses. Chapter 7 of the book describes the creation of a successful optofluidic BFAS adulteration sensor device configuration, which incorporates a geometric flaw. The objective of this setup is to enhance and supervise the impurities in a temperature-dependent ratio of hydrated mono-alcohol fuel, considering various weather conditions. The PPBG has a limited spectral sensitivity profile (SSP) as a result of a geometric flaw in the design of the BFAS sensor device. This defect exhibits a high level of sensitivity to variations in the refractive index (R.I.) of the fuel-filled core during hydration. The newly designed optofluidic BFAS device configuration demonstrates a substantial improvement in P.P. The metrics of the current sensors are superior to those of previous similar sensors, mainly because of the smallest full width at half maximum (FWHM) of the single scattering peak (SSP). The BFAS device configuration has achieved a peak sensitivity of 1057.3 nm/RIU under all weather conditions. The sensitivity of this sensor exceeds that of the surface plasmon resonance optic sensor in detecting hydrated mono-alcohols, but only under static temperature conditions. Reducing the full width at half maximum (FWHM) of the sensor signal profile (SSP) in this configuration improves the detection accuracy (D.A.) and quantification precision (Q.P.), since both metrics are inversely linked to the FWHM.

Chapter 8 focuses on an analysis of a high-contrast optical inline filter that possesses dual electrical adjustment capabilities. This optical inline filter is specifically developed for low-power multiwavelength LASER applications. It is positioned in close proximity to the PMMA cavity, alongside a single-crystal ferroelectric PMN-PT. An efficiency of almost 90% has been attained in the photonic bandgap (PBG) under investigation by manipulating the angle of incidence and applying direct current (DC) voltages, resulting in the transmission peak via this structure. To regulate the thickness of the cavity, it is important to provide an external voltage to the PMN-PT single crystal. Consequently, the PBG structure may be modified correspondingly. This technology has a broad range of application, spanning up to 129 nm. This is achieved by utilizing the reversible domain switching capabilities of PMN-PT layers, which requires a little voltage modification of around 20 V. Hence, the optical inline filter created by BFs shows great potential for various multiwavelength LASER applications, fulfilling the demands of present-day situations such as optical instrument testing, optical communications, fiber sensor systems, optical signal processing, modern telecommunications, and sensors multiplexing industry.

In Chap. 9, we have examined the non-linear transmission of optical pulses with a duration of several hundred femtoseconds in solid-core Bragg fibers. An investigation was conducted on the impact of pulse-width, peak power, soliton order, and higher order dispersion on the evolution of sech pulses in the nonlinear regime. Theoretical models indicate that the early widening is mostly influenced by the effects of group velocity dispersion (GVD) and self-phase modulation (SPM), whereas the creation of dispersive waves becomes apparent after a specific length of the fiber. The solid-core Bragg fibers (also known as DDBFs) can be created using the widely used MCVD technology. This technology is both reproducible and economically viable once it is standardized. It provides a suitable platform for studying nonlinear optical effects at sub-micrometer wavelengths, taking into account various dispersion characteristics.

Samastipur, Bihar, India Ritesh Kumar Chourasia
Pune, Maharashtra, India Aavishkar Katti

Acknowledgment

I begin by extending my sincere gratitude to the Goddess "Shyama" and God "Mahadev" (Kashi Vishwanath) for the insight and tenacity they bestowed upon me over the drafting period of our research monograph book. I sincerely ask for the blessing of Late Pt. Lalit Narayan Mishra, a magnificent human being who founded L.N.M.U., Darbhanga, and is known for his efforts in education, and for giving me the chance to pursue book writing. I humbly like to acknowledge the Publishing Editor, Dr. Sam Harrison, and the Project Coordinator, Dear Mr. Ragavendar M., for their continuous faith in us, and kind support throughout the monograph writing and submission process.

Everything results from group efforts, and this job is no different. I want to take this chance to thank everyone for their assistance and cooperation. First and foremost, I want to express my gratitude to my parents for providing me with all the resources I need to advance in my life. Since I was a young kid, my father, Shri Ashok Chourasia, has emphasized the importance of my character development via study and instilled in me the love of intellectual pursuit. I was truly nurtured by my mother, Smt. Madhavi Chourasia, who loved me tenderly and with care. Only their unwavering support and love throughout my whole life have inspired me to "be" where I am now. Mom and Dad, I appreciate your tolerance and kind assistance during my academic and social journey to date. I have the opportunity to live, work, and relax with a close friend (now my wife), which makes my life both more productive and enjoyable. I would like to thank my dearest wife, a young and dynamic Physicist as well as excellent theoretical and experimental researcher Dr. Ankita Srivastava, for long subject discussions, her morale boost, selfless love, care, and continuous support during the inevitable ups and downs in life. Thank You "Ruby." I am also very thankful to my younger brother, my very close friend, and a young theoretical and experimental Physicist Dr. Nitesh K. Chourasia (including his wife Ms. Anjali Chourasia and lovable daughter "Nivanshi Beta") for expert guidance, endless discussions during the book draft, most importantly to wear responsibilities in my absence at home. Thank You "Babu." I am very thankful to my younger sister Ms. Jyotsana Chaurasia (a young environmentalist and entrepreneur) and her husband Mr. Rahul Chaurasia (an energetic team leader) for their care, selfless love, and technical supports whenever I felt helpless. Both of you were always stand with me in every situation. I and our parents are also very thankful to both of you for the precious gift as your charming, intelligent baby "Riddhi Chaurasia." We all love her. Riddhi has poured happiness in our family, and I personally felt boost up with her every smile. Thank you, "Riddhi Beta,"...!

I hereby acknowledge my gratitude to Institute of Science, Department of Physics, Banaras Hindu University, Varanasi and founder Pt. Madan Mohan Malviya Ji for building good characteristics and human values in me. I bestow my heartfelt sense of gratitude and appreciation to my research supervisor Dr. Vivek Singh for his valuable guidance, keen interest, and consistent support extended to me during the course of investigation and preparation of the book. He not only guided me but also was readily available for me and always responded to the draft of each chapter of my book patiently and quickly. His extensive knowledge of the subject, views, and suggestions were always perceptive, helpful, and appropriate. He is an eloquent speaker and a charismatic person with an abundance of ideas.

My acknowledgment would be lacking if I did not extend a special thank you to my younger, lovable, and bright Ph.D. Scholars Mr. Narendra Bihari (Photonics Specialist), Mr. Atish Kumar Sharma (Tandem Photovoltaics Specialist), Mr. Prakash Jha (Advanced Machine Learning energy storage technologies), Mr. Rakesh Kumar (Hydrogen-Storage and DFT Specialist), and Mr. Maneshwar K. Rai (DFT Specialist) for their full-time support during this journey even their busy Ph.D. work assignments.

Finally, I want to express my gratitude to everyone who had trust in me but whose names I was unable to reveal here.

(Dr. Ritesh Kumar Chourasia)

Man is but a speck in the vastness of the Universe and hence I begin by invoking the name of the Lord Almighty who has blessed me immensely throughout the journey of life in every aspect.

The writing of this monograph has been a journey in which many people have helped in some way or the other. Even a word of encouragement in difficult circumstances can go a long way in ensuring continuity in academic success. I would firstly like to acknowledge our Publishing Editor, Dr. Sam Harrison, and the Project Coordinator, Mr. Ragavendar M., for their continuous support throughout the manuscript writing and submission process.

I would like to mention my special and sincere thanks to Prof. Anurag Sharma, Professor, Department of Physics, Indian Institute of Technology, Delhi, for his constant support and encouragement throughout. I would like to acknowledge fruitful discussions with Prof. C. P. Katti, Professor, School of Computer and System Sciences, Jawaharlal Nehru University, New Delhi. He has been instrumental in shaping the understanding of the mathematical foundation of many physical concepts. Various discussions with him on diverse concepts of Mathematical

Physics have immensely helped in the writing of this treatise. Prof. Vijaya Katti, Professor, Indian Institute of Foreign Trade, New Delhi, has provided invaluable support and prevented wavering of my mind in times of difficulty. Last but not least, my wife Ms. Apurva Sharma must be acknowledged for her constant, unflinching support without which writing this book would not have been possible.

(Dr. Aavishkar Katti)

Contents

Chapter 1
Introduction to Bragg Fiber

Abstract The recent advancements in integrating different materials with distinct optical, electronic, and thermomechanical properties into a single multilayer fiber (known as Bragg Fibers), drawn from a preform, are leading to the development of a new generation of multi-material multilayered fibers. These Bragg Fibers (BFs) possess exceptional functionalities that can be achieved at the length scales and costs associated with optical fibers. Various distinctive devices have been created so far in the shape of fiber utilizing this approach, including transversely emitting fiber lasers, fibers with the ability to sense light, heat, or sound on their outer surfaces simultaneously, fibers with crystalline semiconductor cores, narrow band inline filters, optical d-mux, temperature sensors, strain sensors, and various bio-chemo sensors. The use of these BFs in the next fabrics will result in textiles with advanced functionality. Incorporating many materials in BFs optics has effectively resolved persistent challenges encountered in conventional optical fiber applications. For instance, hollow-core all-solid-cladding fibers have successfully utilized photonic bandgap guiding to overcome concerns related to conventional optical fibers. Similarly, soft-glass mid-infrared fibers have been strengthened mechanically to enhance their durability. We provide in this chapter, an overview of the latest advancements in this emerging subject, which is seeing tremendous growth. Additionally, we emphasize the areas where further expansion is expected. Moreover, the findings derived from this research indicate novel approaches in which the fiber drawing process might be utilized as a manufacturing methodology. Specifically, we discuss current endeavors, natural inspirations, and historical developments to date of multi-material multilayered Bragg fibers in subsequent sections of the chapter.

1.1 Introduction (Natural Inspiration and Historical Development)

Hollow-core multilayer and microstructured fibers, commonly referred to as Bragg Fibers (BFs), have garnered significant interest in recent times. This is primarily due to their ability to effectively guide electromagnetic waves (EM) in the near- and mid-infrared (IR) regions. Additionally, BFs offer several notable advantages over

R. K. Chourasia, A. Katti, *Bragg Fibers*,
https://doi.org/10.1007/978-3-031-65164-9_1

conventional optical fibers (COFs), which are solid in nature. From a fabrication and characterization standpoint, the telecommunication industry has refined the development of COFs to a highly precise and intricate process [1, 2]. Nevertheless, an unforeseen disadvantage of this remarkable achievement has been the concentration of endeavors on a limited range of materials and structures that facilitate the steering of EM waves in the fiber core via total internal reflection within the visible spectrum of silica glass. The intricate scenario has undergone a transformation over the course of a few decades. Microstructured fibers (MFs) have been extensively investigated, enabling a wider range of fiber designs and utilizing the phenomenon of Bragg reflection for electromagnetic (EM) wave guidance [3, 4]. The active pursuit of research in the field of fiber structures is driven by their ease of fabrication and noncomplex nature. This is because the development of new manufacturing technologies has the potential to unlock unique niche applications. The primary benefit of hollow-core Bragg fibers (BFs) in comparison to solid-core conventional optical fibers (COFs) lies in their capacity to confine and direct electromagnetic (EM) radiation mostly within the hollow core. This characteristic allows for the potential reduction in EM wave propagation loss, which may occur owing to material absorption in the fiber core and cladding materials [5–7]. The significance of this phenomenon is notably heightened in the mid- and far-infrared regions, as a majority of materials exhibit substantial levels of absorption in these wavelengths.

Moreover, gaining a comprehensive understanding of the intricate interplay between the shape, composition, and optical properties of biological entities will potentially provide significant inspiration for the development of advanced artificial photonic systems. Design templates for optical technologies are developed by the examination and analysis of natural phenomena. The primary rationale for the development of these novel photonic fiber waveguides is rooted in an analysis of the hierarchical photonic structures observed in the seed coat of *Margaritaria nobilis* fruits [8–10]. The pigmentation of the fruit is a consequence of the phenomenon of light interference occurring within the cellular structures comprising the outer tissue layers of the seed. The aforementioned natural structure exhibits a symbiotic relationship between two components that can be harnessed for the manipulation of light and color [9, 10]. The phenomenon of light scattering occurs when light interacts with structures that possess both micro-scale cylindrical symmetry and nanoscale periodicity, causing the light to disperse in different directions based on its wavelength. This serves as the foundation for groundbreaking bio-inspired photonic fibers, which have the potential to mitigate the pronounced directional chromaticity fluctuations commonly observed in flat multilayer photonic reflectors. The fibers possess the optical property of spectrum filtering and exhibit the color brightness characteristic of cylindrical Bragg stacks. Transparent and elastic synthetic photonic materials offer the potential for constructing multilayer interference fibers that exhibit high transmittance and can be modulated by a transverse mechanical strain. The inception of groundbreaking fiber-based, flexible photonic materials and textiles, capable of adjusting hues across the visible and near-infrared spectrum, is initiated through the design and fabrication of these photonic fibers, drawing inspiration from biological processes.

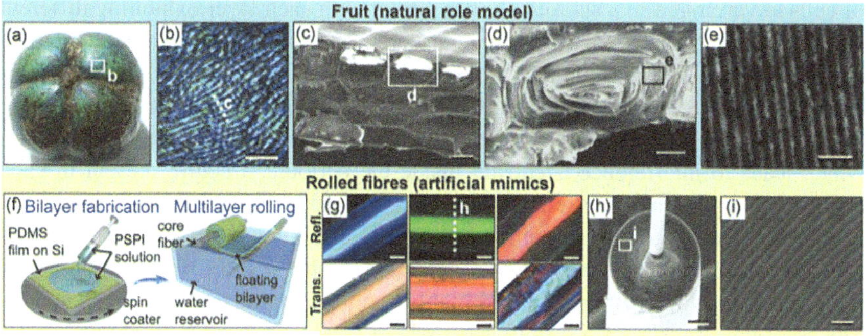

Fig. 1.1 Illustrates the relationship between fruit and fibers. (**a**) The fruit of *Margaritaria nobilis*, lacking its capsule, measures around 10 mm in diameter. (**b**) An optical micrograph depicting the surface of the fruit, reveals the presence of elongated blue cells. (**c**) A scanning electron micrograph depicting a cross-sectional view of the outer layers of the fruit's endocarp reveals the presence of many cell stacks. (**d**) A cross-sectional analysis of a singular tissue cell exposes its internal arrangement, which consists of a concentric, flattened cylindrical layered structure. (**e**) A segment of the stratified architecture within an individual cell is visualized using transmission electron microscopy. (**f**) A schematic diagram illustrating the process of creating artificial photonic fibers. (**g**) Optical micrographs depicting three rolled-up multilayer fibers are presented, showcasing variations in layer thicknesses and colors in both reflection (top) and transmission (bottom). (**h**) A scanning electron micrograph is presented, depicting the cross-section of a fiber. The micrograph reveals the presence of a multilayer cladding that is tightly wrapped around the glass core of the fiber. (**i**) A scanning electron micrograph depicting the distinct layers present in the cladding. The scale bars provided in the image are as follows: 200 μm (**b**), 20 μm (**c**), 10 μm (**d**), 500 nm (**e**), 20 μm (**g**), 20 μm (**h**), and 1 μm (**i**). (Reproduced with permission from [16])

The vibrant hues, exceptional clarity, intense whiteness, and profound darkness observed in natural phenomena are contingent upon the presence of organized, semiorganized, or disorganized arrangements characterized by lattice constants or scattering element sizes that are approximately equivalent to the wavelength of visible light. The spectral composition of reflected and transmitted light is significantly modified in various creatures due to the presence of diverse biological photonic structures, which induce interference or diffraction. This phenomenon gives rise to the visually striking structural colours observed in these organisms [11, 12]. The utilization of one-dimensional multilayer arrangements holds significant significance in the generation of structural colors observed in natural systems, with particular emphasis on their examination within the realm of insects [12, 13]. There has been a growing body of literature documenting the presence of layered photonic systems in different plant species [13, 14]. The fruits of the plant *Margaritaria nobilis*, found in the rainforests of Middle and South America as well as Southern Asia, exhibit a visually remarkable blue-green coloration (Fig. 1.1a) [14–16]. The plant has a partial dependence on avian-mediated seed dissemination, as shown by the potential attraction of birds to its vibrant visual presentation [12–16]. The cellular structure of the fruit's blue seed coat is characterized by elongated cells that predominantly exhibit shades of blue or green (Fig. 1.1b). Multiple layers of cells are arranged in a stacked configuration, with each layer exhibiting different cylindrical orientations of the individual cells (Fig. 1.1c). The examination of a single cell's cross-section indicates that its internal volume is entirely filled with a concentrically layered structure that exhibits periodicity with an approximate value of 200 nm (Fig. 1.1d, e) [16]. The fruit surface is subject to additional illumination, which leads to interference occurring inside the periodic structure present in each cell. As a consequence, the reflection of blue light takes place.

When subjected to directional illumination, a multilayer interference structure is capable of exhibiting its brilliant correlation solely in the specular reflection direction. The coloration of Bragg stacks is highly contingent upon the angle of incidence. The color blue undergoes a shift towards higher wavelengths when examined under diffuse illumination as the observation angle increases. The wide range of angles found in *M. nobilis* fruits can be explained by the presence of a little curvature at the microscopic level, which is combined with the regularity of the layered structure at the nanoscale inside each cell of the seed coat tissue. Additionally, the general curvature of the fruit contributes to this phenomenon [16]. When subjected to directed illumination, a significant portion of the curved multilayer within most individual cells is aligned in a manner that fulfills the specular reflection condition. This alignment results in a visually distinct, pixelated sparkle of changing hues, which is contingent upon the angle at which light strikes the surface at different locations. The glitter in diffuse light is diminished because the light reflected by each individual cell of the fruit towards the observer comes from light that has been incident on the fruit from multiple angles. This results in a more uniform color that gradually transitions over the fruit due to its overall curved shape.

The notion of seeing the creation of distinct structural and optical features through the integration of structures at various length scales within hierarchical synergistic assemblies is frequently found in the study of natural systems [17]. The aforementioned principle is similarly employed in novel optical technologies [18–28]. The seed coat of the *M. nobilis* fruit possesses a hierarchical photonic architecture that plays a crucial role in generating vibrant blue and green colors (Fig. 1.1a). The aforementioned phenomenon serves as a source of inspiration for the development of synthetic photonic fibers that possess many layers and exhibit optical properties that are determined by the intricate combination of nanoscale regularity and overlaid microstructural curvature. The primary objective of this study is to examine the artificially bio-inspired Bragg fiber system, which comprises concentric multilayer wrapped fibers. The system is characterized by a radially periodic refractive index profile that is constructed using only two alternating phases. Throughout this book, the discussion and analysis will be centered on this specific topic. The artificial system exhibits similarities to its natural counterpart in terms of dimensions and underlying optical interactions. Nevertheless, its purpose is to bypass certain complexities found in the inherent arrangement, such as the oval shape of the cells in the fruit's cross-section and any existing fine structure inside the recurring layers.

The concept of optical fiber systems with multilayer claddings was initially proposed by Yeh and Yariv in the late 1970s. Subsequently, advancements in manufacturing techniques have enabled the production of such systems through a normal fiber drawing process. The development of these systems involves the usual extraction of fibers from small preforms [29–36]. There exists a certain set of materials that are prohibited from being incorporated into multilayer fibers. In order to mitigate the occurrence of fractures at material interfaces during high-temperature processing, it is crucial for the selected material combinations to possess an acceptable refractive index contrast and a commensurate thermal expansion coefficient. The proposed methodology employs an alternating technique to fabricate fibers at ambient temperature using a diverse array of soft organic and inorganic materials. These materials possess a wide variety of optical and mechanical properties and are not limited to the conventional symmetry observed in thermally pulled fibers. The Bragg fiber can be composed of a combination of two copolymers, specifically polydimethylsiloxane (PDMS) and polyisoprene-polystyrene triblock copolymer (PSPI) [16]. These two materials are readily accessible in large amounts within the industrial sector and exhibit a refractive index contrast of around 0.13, as determined through the use of ellipsometry. Bragg fibers are fabricated through an initial process of creating a bilayer structure using the two constituent materials. This bilayer is then carefully rolled onto a thin glass mandrel (Fig. 1.1f), typically with a thickness of approximately 20 micrometers. This rolling process results in the formation of a multilayer cladding for the Bragg fiber. The selected material solutions offer significant benefits in terms of the controlled adjustment of the optical performance of Bragg fibers. Additional dielectric materials that have been employed in the production of fibers encompass thermoplastics, such as polystyrene (PS) and poly-methyl-metha-acrylate (PMMA) [37–40]. The rolling process has been utilized to insert thin metal films into non-stretchable fibers of macroscopic diameters [41]. Spray coating and blade coating are two widely utilized procedures in industrial settings that can be effectively employed in roll-to-roll processing for the manufacturing of bilayers on a bigger scale. The thickness of each film in the initial bilayers can be adjusted during the process of film deposition. As a result, the spectral location of the reflection band of the fibers can be flexibly manipulated. The fibers exhibiting a high level of reflection throughout various color spectrums, together with their matching complimentary colors in transmission, are depicted in Fig. 1.1g. The scanning electron microscopy (SEM) images depict the cross-sectional view of the fiber, revealing the presence of concentric multilayer cladding consisting of 80 periods that are tightly wound around the glass mandrel. This is illustrated in the accompanying (Fig. 1.1h, i). The Bragg fibers exhibit a notable reflection (Fig. 1.2d, e) within a specific range of wavelengths due to their multilayer periodicity, similar to the optical characteristics observed in tropical fruit (Fig. 1.2a, b). This reflection is accompanied by a decrease in transmission (Fig. 1.2f, g). Bragg fibers, which are coated with many layers of claddings spanning up to 150 periods, have a reflectivity exceeding 90% throughout their reflection band when employing lower contrast polymer pairings. The bandwidth of these fibers ranges from 70 nm to 30 nm, with a diminishing trend as the number of layers in the cladding increases. The acquisition of spectral data along a line that is perpendicular to the axis of the fruit cell (as depicted in Fig. 1.2a) enables the reconstruction of the spatial distribution of reflection intensity (Fig. 1.2c), which exhibits axial symmetry. The

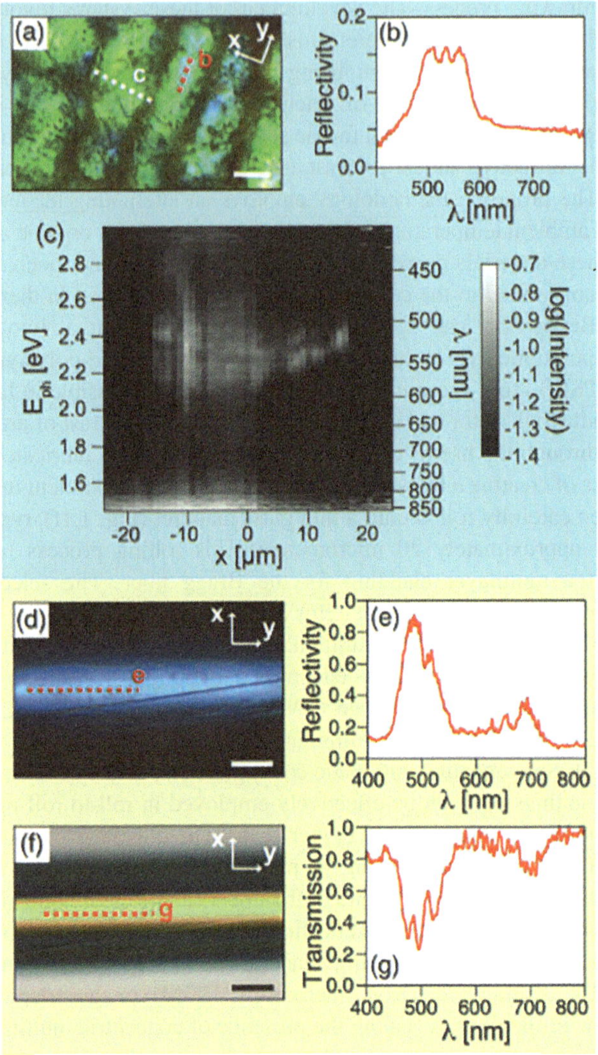

Fig. 1.2 Presents the optical characterization of both the fruit and the fibers. (**a**) The provided image is an optical micrograph that depicts the surface of a fruit, specifically highlighting the presence of elongated tissue cells. (**b**) The reflection spectrum that corresponds to the red dashed line in (**a**) was averaged. (**c**) The spectrally resolved intensity distribution obtained from a line scan (shown by a white dashed line in (**a**)) that traverses a single cell radially, demonstrates the changes in the reflection band caused by the curved structure of the layered nanoscale architecture. In this context, E_{ph} represents the energy associated with a photon, whereas x-denotes the spatial position radially around the fiber, with the fiber axis serving as the central reference point. Figure (**d**) displays an optical micrograph depicting a blue fiber observed under reflection. The reflection spectrum that corresponds to the red dashed line in (**c**) was averaged. Figure (**f**) displays an optical picture of the identical fiber in a state of transmission. The transmission spectrum that corresponds to the red dashed line in (e) was averaged. The scale bars for all measurements are set at a length of 20 micrometers (μm). (Reproduced with permission from [16])

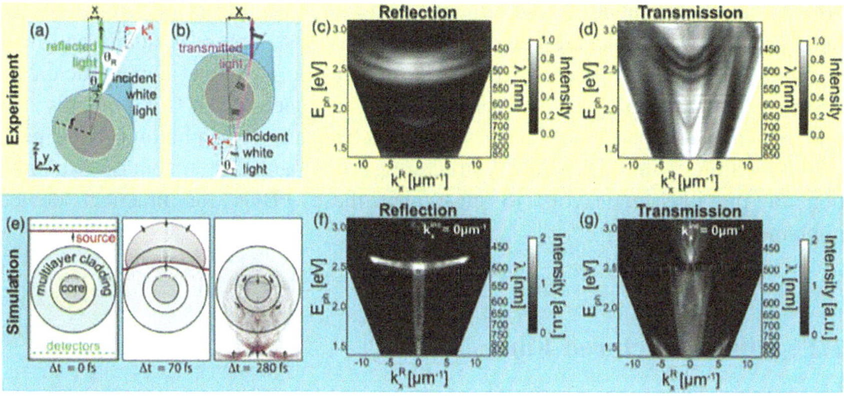

Fig. 1.3 Illustrates the comparison between optical modeling and experimental data. (**a**) This figure presents a simplified diagram illustrating the reflection of light from a fiber in the vertical direction. It depicts the relationship between the projected radial distance (x) from the symmetry axis of the multilayer fiber and the angle of light incidence (θ_R). Notably, this sketch does not account for the effects caused by the finite numerical aperture of the illumination and collection optics, such as extended light cones. (**b**) A comparable diagram illustrating the transmission of light in a transverse manner through the fiber. (**c**) The provided diagram illustrates the band structure of fiber reflection for the light within the visible spectrum. (**d**) The band diagram illustrating the transmission characteristics of the fiber. (**e**) The image on the left depicts the configuration of the FDTD simulation cell employed for optical modeling. The subsequent pair of photos illustrate the transverse propagation of the light pulse through the multilayer fiber at two subsequent periods. To enhance clarity, the fiber is depicted in its outlines. The logarithm of the light pulse's field amplitudes increases from blue to red in all three photos (**f**, **g**). Band diagrams for reflection and transmission were derived from the simulations for perpendicular light incidence ($k_x = 0\ m^{-1}$) It showcases the cross-sectional view of the fiber, which exhibits an inner cladding periodicity of 710 nm (indicated in yellow) and an outer periodicity of 340 nm (depicted in light blue). The incident light pulse, spanning the entire visible range, is represented by the colors red and blue. Additionally, the image highlights the presence of reflection and transmission detectors, indicated the subsequent pair of photos depict the transverse propagation of the light pulse within the multilayer fiber at two subsequent time intervals. To enhance clarity, the fiber is represented by its outlines. All three photos depict the progressive increase in the logarithm of the field amplitudes of the light pulse, with the color spectrum ranging from blue to red. The reflection and transmission band diagrams for perpendicular light incidence ($k_x = 0;\ m^{-1}$) were derived from the simulations. (Reproduced with permission from [16])

aforementioned measurement was conducted on the fibers to ascertain the distribution of their reflection and transmission intensities. The concentric multilayer reflector, due to its curved structure, can convert the lateral distance from the symmetry axis of the fiber in each collected spectrum into a corresponding incidence angle. Please refer to Fig. 1.3a, b for visual representation. This allows for the direct measurement of the dispersion relation of the Bragg fiber system in both reflection and transmission, as shown in Fig. 1.3c–g.

Moreover, the incorporation of a defect layer into a BFs structure enables the creation of a pass band within a photonic band gap (PBG). This capability empowers researchers to construct BFs that offer precise manipulation of the wavelengths and orientations of propagating electromagnetic waves. The impact of semiconductor materials on several societal domains in the twenty-first century is significant. BFs, in particular, have the potential to play a pivotal role in optical communications, current sensor technologies, and the inline optical filter sectors. Consequently, the primary chapter of the book centers on the advancement of photonic band gaps (PBGs) within cylindrical periodic structures, accompanied by a concise examination of previous endeavors to customize these PBGs for practical device implementations.

1.2 Microstructured Fibers (MSFs)

The refractive index (R.I.) profile, which is determined by the composition of the optical materials, plays a crucial role in controlling the light propagation and distinctive optical characteristics observed in conventional optical fibers. The design freedom of the optical qualities in this particular case is limited by the properties of the material. It is implausible, for instance, to shift the region of anomalous dispersion to wavelengths that are lower than the zero-dispersion value of silica. In the case of micro-structured or photonic crystal fibers (PCF), the optical properties are determined by the dimensions, arrangement, and morphology of the air voids that traverse the entire length of the fiber, in contrast to conventional fibers. The design of optical attributes can be achieved with considerable flexibility due to the ability to manipulate geometrical elements. It is feasible to shift the zero-dispersion wavelength towards the visible spectral region or modify the numerical aperture and mode field area across a wide range. Hollow-core fibers, also known as photonic band gap fibers, facilitate the transmission of higher intensities while reducing the occurrence of nonlinear effects due to the minimized interaction between directed light and the glass material. The unusual optical characteristics of micro-structured fibers (MSFs) have garnered the attention of academics who are exploring their potential as waveguides for future generations. There are two distinct groups of MSFs that have been identified [42].

1.2.1 Solid-Core MSFs

The cladding component of these multimode optical fibers (MSFs) consists of a lattice arrangement of air holes, whilst the core is composed of a solid material. The guidance of optical waves is facilitated by a modified total internal reflection system (Fig. 1.4a), which has a resemblance to the light guidance process observed in classic fiber [43, 44]. The fibers in question are commonly known as photonic crystal fibers or holey fibers (PCFs) (Fig. 1.4b).

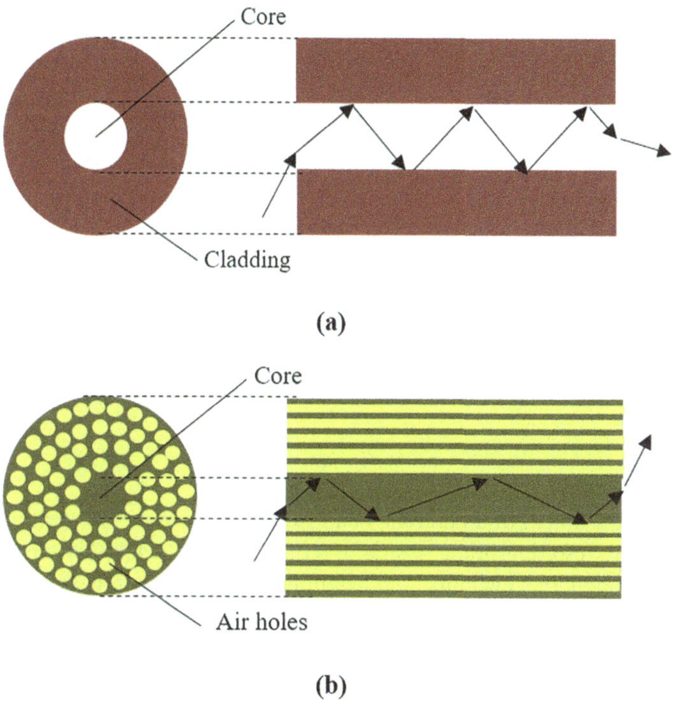

Fig. 1.4 TIR in (**a**) Conventional fiber and (**b**) PCFs

1.2.2 Hollow-Core MSFs

The utilization of total internal reflection for light guidance in hollow-core MSFs is not feasible due to the lower refractive index of the core compared to the effective index of the cladding [28, 42–44]. The presence of a photonic band gap (PBG) in the band diagram at specific wavelengths results in the exclusion of optical waves of those wavelengths from entering the cladding region. This exclusion mechanism facilitates the phenomenon of light guiding in photonic band gap fibers. The band gap refers to a specific range of wavelengths in which the photonic bandgap (PBG) material effectively inhibits the propagation of light, irrespective of its direction and polarization. There exist two distinct types of hollow-core MSFs.

1D-Bragg Fibers (BFs)

The Bragg fiber is characterized by a low-index core positioned at the center, surrounded by a periodic multilayer cladding consisting of concentric layers (Fig. 1.5). These cladding layers are composed of alternating materials with high and low refractive indices, each of which exceeds the refractive index of the core

Fig. 1.5 Design of vacuum core Bragg fiber

[36]. The establishment of a photonic band gap arises from the inherent periodicity of the cladding, resulting in the obstruction of light with certain wavelengths from traversing through the cladding and confining it solely within the core region [45, 46]. The containment of optical waves within the core of an optical fiber is achieved through the utilization of Bragg diffraction induced by the periodic claddings. The utilization of an air or vacuum core within a fiber, provided that the fiber has been appropriately engineered to confine the majority of optical power to the core, is a viable approach. The Bragg fiber exhibits potentially distinctive characteristics, such as ultra-low loss, minimal dispersion, and negligible radiation loss, making it a promising candidate for achieving the ultimate transmission line. Moreover, the implementation of a transmission line that is independent of wavelength is achievable through the utilization of Bragg fiber. The selection of cladding material can be flexibly adjusted according to the wavelength used, as the hollow core is responsible for transmitting the majority of optical power. Bragg fiber designs can also enable single-guided modes that do not exhibit azimuthal dependency. The guided modes in Bragg fibers are single modes, in contrast to the basic mode seen in conventional fibers, which always exhibit double degeneracy. Consequently, numerous polarization-dependent phenomena that are undesirable in Bragg fibers can be eliminated.

2D-Photonic Crystal Bandgap Fibers (PCBFs)

The cladding zone of hollow-core photonic crystal fibers (PCFs) is comprised of the air-hole defect (Fig. 1.6). The defect hole of the hollow-core photonic crystal fiber (PCF) is designed in a manner that enables the propagation of light within it, specifically for wavelengths falling within the crystal band gap [47, 48]. The inability of this light to penetrate the crystal is attributed to its wavelength falling within the band gap. Consequently, the wave becomes trapped within the imperfection [48].

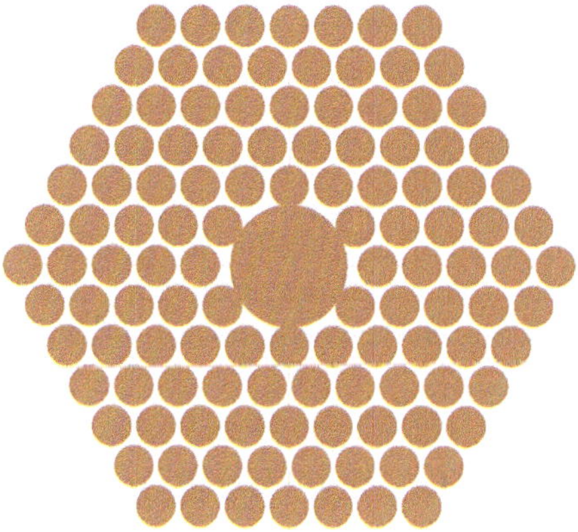

Fig. 1.6 Design of hollow-core 2D-PCF

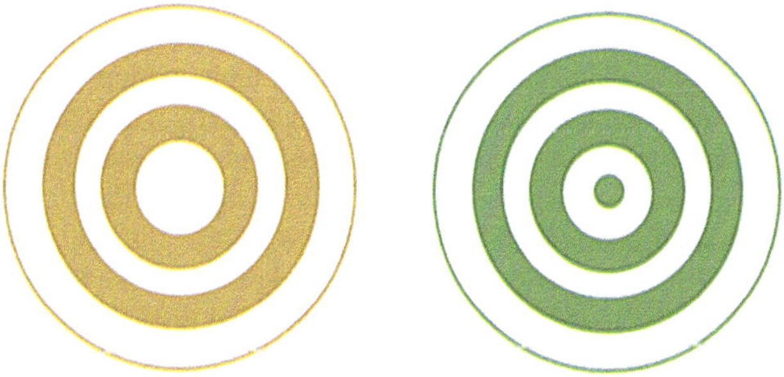

Fig. 1.7 Schematics of air core Bragg fiber and coaxial Bragg fiber

1.3 Varieties of Bragg Fibers

The Bragg fiber structure can be characterized and constructed by several approaches, as it constitutes a periodic layered cylindrical arrangement surrounding the core. The following constructs are a selection of examples.

1.3.1 Omniguide Bragg Fiber

The Omniguide/Coaxial Bragg fibers, characterized by concentric cylindrical multilayer formations with a homogenous low index core, are referred to as such [49]. Coaxial Bragg fibers employ the principle of Bragg reflection to guide light, wherein a smaller high-index column is inserted within the low-index core. The fiber structure depicted in Fig. 1.7 is presented.

1.3.2 High R.I. Core Bragg Fiber

A core possessing a high index defines a Bragg fiber as a micro-structured fiber that exhibits cylindrical symmetry (Fig. 1.8). It is characterized by a core composed of a high-index material, often silica, and a cladding consisting of alternating layers of silica and a material with a lower refractive index. In contrast to conventional Bragg fibers, the silica core of the fiber under consideration offers guidance, similar to photonic crystal fibers that lack a central hole.

1.3.3 Low R.I. Core Bragg Fiber

As depicted in Fig. 1.5, the Bragg fiber under consideration consists of a low-index core encompassed by a multilayer cladding with periodicity (concentric layers). The cladding is composed of alternating materials with high and low refractive indices, both of which exceed the refractive index of the core [36]. The inherent periodicity of the cladding results in the formation of a photonic band gap, which effectively restricts the propagation of light with specific wavelengths to the core region, while allowing it to partially penetrate the cladding [45, 46]. The utilization of Bragg diffraction, induced by the periodic claddings, serves to confine the propagation of optical waves exclusively within the core of the optical fiber. By utilizing an air or vacuum core and implementing appropriate fiber construction techniques, it is possible to confine the bulk of optical power within the core. Consequently, the disparity in refractive index between the adjacent bilayers in consecutive periodic configurations significantly impacts the propagation of waves within these fibers.

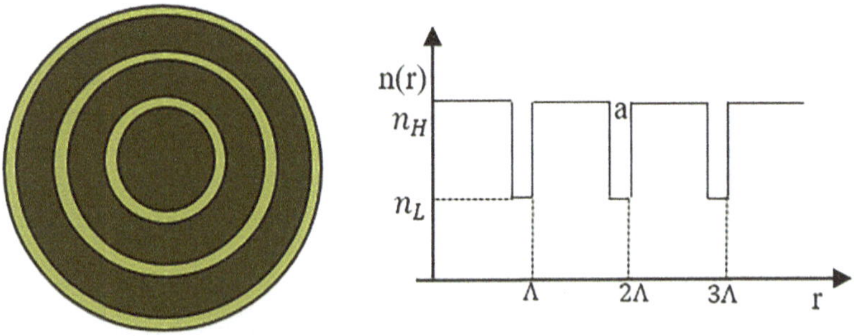

Fig. 1.8 Schematics of a high-index core Bragg fiber along with index profile

1.4 Bragg Fiber Advances and Novel Perspective

In conventional optical fibers, the confinement of EM waves is principally achieved within the high-index central core through the mechanisms of total internal reflection and photon propagation. The Bragg reflection phenomenon presents a unique approach for manipulating photons in comparison to conventional confinement techniques. Yeh et al. [36] were the first to propose the concept of Bragg, so introducing this notion. Nevertheless, the lack of advanced production techniques and questions over its practicality resulted in a significant gap in research in this field for a period exceeding many years [50]. Against the backdrop of an increased focus on photonic crystals throughout the late 1990s, there was a renewed interest in investigating the feasibility of constructing hollow-core fiber with broadband capabilities and low loss. Several studies conducted during this period demonstrated the successful implementation of such fibers across a variety of wavelengths [18, 51, 52]. The presence of several problematic alternatives in Bragg fibers, as opposed to normal fibers, is not surprising due to the various guiding mechanisms employed by Bragg fibers and conventional optical fibers. An example of this phenomenon can be observed in the present upswing in interest in the possibility of manipulating light within an air core [53, 54]. Moreover, Bragg fiber can provide a single-guided mode without azimuthal dependency, specifically in the TE or TM mode. The guided fiber modes under consideration are characterized as single modes, in contrast to the fundamental mode seen in conventional fibers, which consistently exhibit double degeneracy. Consequently, the complete elimination of numerous adverse polarization-dependent effects in Bragg fibers can be achieved [49]. The structure, composition, and optical properties of biological photonic systems have the potential to provide significant inspiration for the advancement of artificial photonic components [13, 55, 56]. The production of design templates for optical technologies is facilitated through the examination of natural photonics [57–62].

Bragg fibers are a subject of significant interest and importance within the field of optical communication and photonics. Photonic crystal fibers (PCFs), often referred to as microstructured fibers, possess exceptional characteristics that render them highly advantageous in a wide range of fields such as telecommunications, sensing, and laser technology. This introductory chapter aims to examine the fundamental concepts behind Bragg fibers, tracing their historical development and elucidating their important properties.

References

1. L.L. Auguste, R. Jindal, J.M. Blondy, J. Marcou, B. Dussardier, G. Monnom, D.B. Ostrowsky, B.P. Pal, K. Thyagarajan, 1800 (ps/nm)/km chromatic dispersion at 1.55 μm in dual concentric core fiber. Electron. Lett. **36**, 1689 (2000)
2. M. Artoni, G.C. La-Rocca, Tunable photonic stop bands in homogeneous absorbing media. Phys. Rev. Lett. **96**, 073905 (2006)

3. R.K. Chourasia, V. Singh, Estimation of photonic band gap in the hollow core cylindrical multilayer structure. Superlattice. Microst. **116**, 191–199 (2018)
4. N.K. Chourasia, A. Srivastava, V. Kumar, R.K. Chourasia, Doubly electrically tuned cylindrical Bragg fiber waveguide inline optical filter for multiwavelength LASER applications. Mater. Today Commun. **25**, 101620 (2020)
5. N. Bihari, A. Srivastava, N.K. Chourasia, R.K. Chourasia, H-polarised EM-wave transport in polymeric-chalcogenide columnar photonic materials. J. Phys. Conf. Ser. **2426**, 012009 (2023)
6. R.K. Chourasia, C.S. Yadav, A. Upadhyay, N.K. Chourasia, V. Singh, Comparative study of defect mode intensity and wavelength modulation in Bragg fiber waveguide sensors. Optik **204**, 164198 (2020)
7. R.K. Chourasia, C.S. Yadav, A. Upadhyay, N.K. Chourasia, V. Singh, Analysis of Bragg fiber waveguides having a defect layer for biosensing application. Optik **200**, 163400 (2020)
8. K. Chung, S. Yu, C.J. Heo, J.W. Shim, S.M. Yang, M.G. Han, H.S. Lee, Y. Jin, S.Y. Lee, N. Park, J.H. Shin, Flexible, angle-independent, structural color reflectors inspired by morpho butterfly wings. Adv. Mater. **24**(18), 2375–2379 (2012). https://doi.org/10.1002/adma.201200521. Epub 2012 Apr 10. PMID: 22489053
9. A. Pris, Y. Utturkar, C. Surman, et al., Towards high-speed imaging of infrared photons with bio-inspired nanoarchitectures. Nat. Photonics **6**, 195–200 (2012). https://doi.org/10.1038/nphoton.2011.355
10. M. Kolle, P. Salgard-Cunha, M. Scherer, et al., Mimicking the colourful wing scale structure of the Papilio blumei butterfly. Nat. Nanotechnol. **5**, 511–515 (2010). https://doi.org/10.1038/nnano.2010.101
11. S. Kinoshita et al., Physics of structural colors. Rep. Prog. Phys. **71**, 076401 (2008)
12. P. Vukusic et al., Brilliant whiteness in ultrathin beetle scales. Science **315**, 348–348 (2007). https://doi.org/10.1126/science.1134666
13. L.P. Biro, J.P. Vigneron, Photonic nano architectures in butterflies and beetles: Valuable sources for bio inspiration. Laser Photonics Rev. **5**, 27 (2011)
14. S. Kinoshita, S. Yoshioka, Structural colors in nature: The role of regularity and irregularity in the structure. Chemphyschem **6**, 1442 (2005)
15. X. Liu, S. Zhang, H. Zhang, Microstructure of butterfly wing scale and simulation of structural color. Optik **127**(4), 1729–1733 (2016)
16. M. Kolle, A. Lethbridge, K. Moritz, J.J. Baumberg, J. Aizenberg, P. Vukusic, Bio-inspired band gap tunable elastic optical multilayer fibers. Adv. Mater. **29**, 2239–2245 (2013)
17. D. Shandon Hart et al., External reflection from omnidirectional dielectric mirror fibers. Science **296**, 510–513 (2002). https://doi.org/10.1126/science.1070050
18. F. Brechet, P. Roy, J. Marcou, D. Pagnoux, Single-mode propagation into depressed-core-index photonic band gap fiber designed for zero-dispersion propagation at short wavelengths. Electron. Lett. **36**, 514 (2000)
19. N.K. Chourasia, A. Srivastava, V. Kumar, R.K. Chourasia, Optimizing temperature-dependent molar volume fraction of biodiesel fuel in a pseudo-binary mixture through Bragg fiber waveguide sensor having defect layer. Fuel **293**, 120489 (2021)
20. N.K. Chourasia, N. Bihari, R.K. Chourasia, An optofluidic Bragg fiber sensor for estimating adulterants in a temperature-dependent molar fraction of hydrated mono-alcohol fuels. Heliyon **8**(9), e10532 (2022)
21. R.K. Chourasia, S. Prasad, V. Singh, A novel Bragg fiber waveguide based narrow band inline optical filter. Optik **169**, 269–276 (2018)
22. M. Campbell, D.N. Sharp, M.T. Harrison, R.G. Denning, A.J. Turberfield, Fabrication of photonic crystals for the visible spectrum by holographic lithography. Nature **404**, 53–56 (2000)
23. R.F. Cregan, B.J. Mangan, J.C. Knight, T.A. Birks, P.S.J. Russell, P.J. Roberts, D.C. Allan, Single-mode photonic band gap guidance of light in air. Science **289**, 415–419 (2000)

24. K. Chung, S. Yu, C.J. Heo, J.W. Shim, S.M. Yang, M.G. Han, H.S. Lee, Y. Lin, Y. Jin, S.Y. Lee, N. Park, J.H. Shin, Flexible, angle-independent, structural color reflectors inspired by morpho butterfly wings. Adv. Mater. **24**, 2375 (2012)
25. R.K. Chourasia, S. Prasad, V. Singh, Performance study of a liquid-core Bragg fiber sensor in the presence of a defect layer. Opto-Electron. Rev. **25**(3), 215–221 (2017)
26. R.K. Chourasia, S. Prasad, V. Singh, Voltage-tunable pass-band in a cylindrical multilayered structure containing PMMA and 0.67 PMN–0.33 PT single crystal as a defect layer. Opt. Quant. Electron. **48**(12), 1–9 (2016)
27. R.K. Chourasia, V. Singh, Cylindrical Bragg waveguide-based temperature sensors, UPCON-2016 IEEE 87-90 (2016)
28. R.F. Cregan, B.J. Mangan, J.C. Knight, T.A. Birks, P.S.J. Russell, P.J. Roberts, Single-mode photonic band gap guidance of light in air. Science **285**, 1537–1539 (1999)
29. A. Yariv, P. Yeh, *Optical Wave in Crystals* (Wiley, 2003)
30. A. Yariv, *Optical Electronics in Modern Communications* (Oxford University Press, Oxford, 1997)
31. P. Yeh, *Optical Waves in Layered Media* (Wiley, 2005)
32. P. Yeh, A. Yariv, Bragg reflection waveguides. Opt. Commun. **19**, 427–430 (1976)
33. P. Yeh, A. Yariv, C.S. Hong, Electromagnetic propagation in periodic stratified media. I. General theory. J. Opt. Soc. Am. **67**, 423–437 (1977)
34. P. Yeh, *Optical Waves in Layered Media* (Wiley, New York, 1988)
35. P. Yeh, *Optical Waves in Layered Media* (Willey, Singapore, 1991)
36. P. Yeh, A. Yariv, E. Marom, Theory of Bragg fiber. J. Opt. Soc. Am. **68**, 1196–1201 (1978)
37. E. Pone, C. Dubois, N. Guo, Y. Gao, A. Dupuis, F. Boismenu, S. Lacroix, M. Skorobogatiy, Drawing of the hollow all-polymer Bragg fibers. Opt. Express **14**, 5838–5852 (2006)
38. R.K. Chourasia, N.K. Chourasia, N. Bihari, Optical Properties of Hollow-Core Bragg Fiber Waveguides, Benthem Science 214–236 (23), (2023)
39. R.K. Chourasia, N.K. Chourasia, A. Srivastava, N. Bihari, Photonic Nanostructured Bragg Fuel Adulteration Sensor, Benthem Science 237–264 (28) (2023)
40. N.K. Chourasia, A. Srivastava, V. Kumar, R.K. Chourasia, Comparative performance study of liquid core cylindrical Bragg fiber waveguide biosensors. Pramana **95**, 9 (2021)
41. N. Gibbons, J.J. Baumberg, C. Bower, M. Kolle, U. teiner, Scalable cylindrical metallodielectric metamaterials. Adv. Mater. **21**, 3933 (2009)
42. F. Poli, A. Cucinitta, S. Selleri, *Photonic Crystal Fibers: Properties and Applications* (Springer, Dordrecht, 2007)
43. J.C. Knight, T.A. Birks, P.S. Russell, D.M. Atkin, All- silica single mode optical fiber with photonic crystal cladding. Opt. Lett. **21**, 1547–1549 (1996)
44. P. Russel, Photonic crystal fibers. Science **299**, 358–362 (2003)
45. S. Dasgupta, B.P. Pal, M.R. Shenoy, Design of dispersion compensating Bragg fiber with an ultrahigh figure of merit. Opt. Lett. **30**, 1917–1919 (2005)
46. S. Dasgupta, B.P. Pal, M.R. Shenoy, Nonlinear spectral broadening in solid-core Bragg fibers. J. Lightwave Technol. **25**, 2475–2481 (2007)
47. F. Zolla, G. Renversez, A. Nicolet, B. Kuhlmey, S. Guenneau, D. Felbacq, *Foundations of Photonic Crystals Fibers* (Word Scientific, 2005)
48. I.A. Sukhoivanov, I.V. Guryev, *Photonic Crystals: Physics and Practical Modeling* (Springer, 2009)
49. M. Ibanescu, Y. Fink, S. Fan, E.L. Thomas, J.D. Joannopoulos, An all dielectric coaxial waveguide. Science **289**, 415–419 (2000)
50. N.J. Doran, K.J. Blow, Cylindrical Bragg fibers: A design and feasibility study for optical communications. J. Lightwave Technol. **1**, 588–590 (1983)
51. Y. Fink, D.J. Ripin, S. Fan, C. Chen, J.D. Joannopoulous, E.L. Thomas, Guiding optical light in air using an all-dielectric structure. J. Lightwave Technol. **17**, 2039–2041 (1999)
52. Y. Fink, J.N. Winn, S. Fan, C. Chen, J. Michel, J.D. Joannopoulos, E.L. Thomas, A dielectric omnidirectional reflector. Science **282**, 1679–1682 (1998)

53. N. Croitoru, J. Dror, I. Gannot, Characterization of hollow fibers for the transmission of infrared radiation. Appl. Opt. **29**, 1805–1809 (1990)
54. M. Miyagi, A. Hongo, Y. Aizawa, S. Kawakami, Fabrication of germanium coated nickel hollow waveguides for infrared transmission. Appl. Phys. Lett. **43**, 430–432 (1983)
55. L. Lee, R. Szema, Inspirations from biological optics for advanced photonic systems. Science **310**, 1148–1150 (2005)
56. A. Parker, H. Townley, Biomimetic of photonic nanostructures. Nat. Nanotechnol. **2**, 347–353 (2007)
57. B.T. Hallam, A.G. Hiorns, P. Vukusic, Developing optical efficiency through optimized coating structure: Biomimetic inspiration from white beetles. Appl. Opt. **48**, 3243–3249 (2009)
58. J.P. Vigneron, M. Rassart, C. Vandenbem, V. Lousse, O. Deparis, L.P. Biro, D. Dedouaire, A. Cornet, P. Defrance, Spectral filtering of visible light by the cuticle of metallic woodboring beetles and microfabrication of matching bioinspired material. Phys. Rev. E **73**, 041905 (2006)
59. R. Potyralio, H. Ghiradella, A. Vertiatchikh, K. Dovidenko, J.R. Cournoyer, E. Olson, Morpho butterfly wing scales demonstrate highly selective vapour response. Nat. Photonics **1**, 123–128 (2007)
60. A.D. Pris, Y. Utturkar, C. Surman, W.G. Morris, A. Vert, S. Zalyubovskiy, T. Deng, H.T. Ghiradelia, R.A. Potyrailo, Towards high-speed imaging of infrared photons with bio-inspired nano architectures. Nat. Photonics **6**, 195–200 (2012)
61. R.K. Chourasia, V. Singh, *Comparative analysis of the dispersion characteristic of Bragg fibers having different refractive index contrast*, in International Conference on Fiber Optics and Photonics, Th3A. 75 (2016)
62. N. Bihari, N.K. Chourasia, R.K. Chourasia, Exploring optical properties in cylindrical polymeric-chalcogenides photonic materials. Mater. Today Proc. **67**(5), 625–631 (2022)

Chapter 2
Multilayered and Multi-material Fabrication Techniques and Detailed Processes for Bragg Fibers

Abstract Conventional solid-core optical fibers need the utilization of materials with outstanding transparency. The identification of such materials has proven to be difficult due to the intrinsic limitations associated with the transmission of light through solids, such as absorption, scattering, and nonlinear processes. Hollow optical fibers have the capacity to diminish the dependence on the transparency of fiber materials for the transmission of light. The chapter offers a comprehensive examination of the design and illustration of a hollow optical fiber that is covered with inner omnidirectional multilayer reflectors. The significant photonic bandgaps are accountable for the confinement of light within the hollow core. Recent research has demonstrated notable progress in combining many materials with unique optical, electrical, and thermomechanical characteristics into a single multilayer fiber. This procedure entails extracting the fiber from a preform. The result is the creation of a novel fabric that incorporates different elements and exhibits distinctive capabilities. These capabilities can be accomplished using the size and expenses related to optical fibers. Using this technique, many new fiber-based devices have been made, including transversely emitting fiber lasers, fibers with sensory capabilities for vision and hearing, surface-emitting applications, fibers that can sense light, heat, or sound on their exterior, and fibers with crystalline semiconductor cores. By incorporating these fibers into future fabrics, the resulting textiles will possess enhanced utility. The utilization of several materials in fiber optics has successfully addressed enduring difficulties experienced in traditional applications. For example, multi-material fibers have effectively incorporated photonic bandgap navigation in hollow-core all-solid-cladding fibers and improved the mechanical strength of soft-glass mid-infrared fibers. We offer a comprehensive analysis of the most recent developments in this burgeoning industry and highlight the specific sectors poised for future expansion. The techniques or foundations presented in this chapter will serve as the essential underpinning for the next chapters in this book. Furthermore, the results obtained from this study suggest possible uses of the fiber drawing method as a technology for production. In this discussion, we specifically focus on the ongoing efforts to utilize multi-material-fiber drawing for chemical synthesis and the manufacturing of nanostructures, including arrays of nanowires and structured nanoparticles.

R. K. Chourasia, A. Katti, *Bragg Fibers*,
https://doi.org/10.1007/978-3-031-65164-9_2

2.1 Introduction

Optical fibers have made a big difference in people's lives around the world and have changed the human situation in many important and subtle ways. At present, optical fibers serve as the main method for transmitting the internet and other services globally [1–4]. Furthermore, optical fibers are utilized in a wide range of applications, including noninvasive medical treatments [5] and the monitoring of structural integrity in bridges and oil pipelines [6, 7], biosensors [8–11], adulteration sensors [12–14], temperature sensors [15], and narrowband inline filters [16]. Furthermore, fiber lasers are currently employed in the realm of materials processing and manufacturing [17]. An outstanding characteristic of these enhancements is the remarkable straightforwardness of optical fibers in relation to their material composition. Most optical fibers now in use consist of a single substance called silica glass. Although optical fibers constructed from different glasses or polymers have been created, silica glass continues to be the primary material used in manufacturing optical fibers. During the mid-1990s, researchers successfully created novel optical fibers known as photonic crystal fibers (PCFs) and photonic band gap (PBG) fibers. These fibers were developed using the ideas of photonic band gaps, which were introduced as a novel field in optical physics [18, 19]. These advancements have revealed the clear constraints of the structures that can be created using the conventional method of thermal fiber drawing. The first proposal to produce silica PBG fibers by including air gaps, without the need for any extra material, was deemed impracticable by experienced practitioners. Nevertheless, the rapid success of PBG fibers has led to unexpected consequences for the process of producing fibers. A significant consequence that occurred over the past decade was the development of the concept of "multi-material fibers." This innovative category of fibers utilizes the capabilities of conventional fiber production methods, with a particular focus on creating novel fiber shapes, functionalities, and applications by modifying the material composition of the fiber. It is difficult to give a clear and complete definition of this evolving concept. This study presents a comprehensive description of multi-material fibers, which are structures characterized by a high aspect ratio and composed of many distinct materials. These fibers are often produced using thermal drawing from a larger-scale model called a "preform." The fiber format has successfully included a range of materials with unique optical, electrical, thermomechanical, and acoustic properties [20]. The burgeoning field of manufacturing is implementing innovative techniques that frequently do not correlate with optical fibers. Illustrative scenarios encompass fibers that exhibit electroluminescence upon exposure to light on the outer surface [21–23] or in response to temperature fluctuations in the surrounding environment [24]. Additionally, there are fibers capable of self-monitoring their functionality and structural integrity [25]. Furthermore, there exist flexible and lightweight arrays of fibers that capture images of the immediate surroundings [26]. Lastly, there are fibers designed to generate or detect sound or ultrasonic signals [27, 28]. Fiber drawing is progressing as a manufacturing technique to produce electrical and optoelectronic fiber devices. These devices possess the capacity to be incorporated or interlaced with fabrics, augmenting them with sophisticated and inventive functionalities. The emergence of multi-material fibers has led to the development of new insights into the process of fiber drawing. Nanostructures, such as nanowires, may be fabricated by dimensional reduction [29–31], resulting in diameters of a few nanometers and remarkable lengths. The fiber cladding can be viewed as a vessel for chemical synthesis, where the reactive chemicals are located in the core. Furthermore, the utilization of multi-material fibers presents a novel prospect for conducting comprehensive investigations on fluid dynamics in confined spaces and throughout a wide range of dimensions [32]. There have been recent developments in manufacturing technology that can create intricate particles of different sizes, ranging from micro- to nanoscales. This method is capable of being scaled up and makes use of in-fiber technology [33].

Multi-material fibers are an emerging discipline that involves the synthesis of two distinct fiber constituents. Typically, the process begins by preparing a large-scale preform. Given the preform's diameter in the centimeter range, it is feasible to create a complex transverse cross-section by carefully positioning structures composed of many materials. In addition, thermal fiber drawing has an inherent capacity for expansion, allowing for the creation of large quantities of fiber across long distances while keeping exact control over size and uniformity throughout the length. The manufacturing process imposes limitations on the permissible material combinations employed so as to preserve the intricate transverse structure of the multi-material preform. Additional variations exist for this general approach. One instance of additive manufacturing involves the utilization of a single-material fiber or filament to support the production of multi-material fibers. This objective can be accomplished through two methods: dip-coating, which involves coating the outer surface of long fibers; or by introducing vapor into the voids of short segments of an existing fiber.

Despite being in its nascent phase, there have been notable recent advancements in this field. Thus, this research aims to provide a succinct and precise overview of the significant achievements, as per our previously stated concept of multi-material fibers. This study aims to highlight the emerging field that integrates optics, materials science, device physics, nanotechnology, and fluid dynamics. The aim is to inspire more researchers to participate in this subject. Section 2.2 examines "Material Constraints and Fiber Drawing" and explores the restrictions of the materials that may be used in the same fiber. Section 2.3 thereafter outlines general techniques for producing multi-material preforms by physical processes. Section 2.3 offers an in-depth analysis of several physical methods for creating multi-material fibers that include sophisticated features, including electrical, optoelectronic, and piezoelectric capabilities. The approaches included in this list are the Rod-in-tube method, Extrusion, Stack-and-Draw method, and thin-film rolling techniques. The talk also examines the manufacturing of fibers including conventional crystalline semiconductor cores. Section 2.4, headed "Deep-coating approach based on chemical route approaches," explores novel ways that utilize fiber drawing to accomplish chemical synthesis within the fiber. Further, Sect. 2.5 presents a comprehensive review of multi-material fibers which are further utilized for the purpose of guiding light along the fiber's axis. In Sect. 2.6, titled "Future Prospects," we provide a summary of the present chapter by examining the production techniques used for hollow-core multi-material PBG fibers and exploring possible future advancements.

2.2 The Material Limitations and the Process of Fiber Drawing

Optical fibers are produced by thermally pulling them from a bigger preform in a fiber draw tower. The preform-to-fiber process, seen in Fig. 2.1 [34], is employed in this procedure. The preform is inserted into a furnace and subjected to heat until it reaches a temperature at which the material undergoes a transition to a pliable state. Afterward, either gravity or an external force is used to extract the molten gob from the tip of the preform, resulting in its elongation into a thin strand. The gauge is used to measure the diameter of this thread. If necessary, polymer coatings can be added to the fiber surface to provide mechanical protection. The manufacturing process employed for optical fiber production is quite simple and yields a vast network of communication cables spanning thousands of kilometers globally [1, 2]. On the other hand, the double-crucible method [35] has been utilized to produce optical

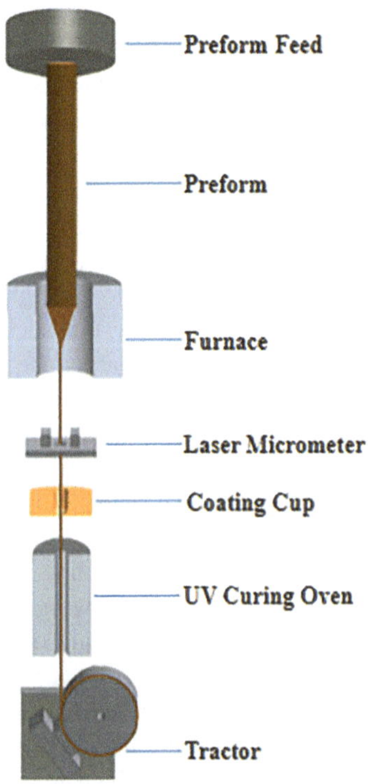

Fig. 2.1 Schematic of the thermal fiber drawing process in a draw tower. (Reprinted from [39])

fibers by forcing molten glass off the core and cladding via a nozzle. Thus yet, there is a lack of recorded instances of multi-material fibers produced utilizing this method.

The characteristics of the draw, such as the speed at which the material is drawn, are influenced by the viscosity of the preform material at the drawing temperature, assuming that the preform is made up of only one material. Conversely, multi-material preforms may comprise constituents, such as metals or crystalline semiconductors, which are incapable of being thermally drawn individually. The utilization of thermal drawing restricts the range of material combinations that arc feasible for implementation in this manufacturing process. The chosen materials encompass three distinct material categories from an electronic perspective: an amorphous insulator, a crystalline semiconductor, and a metal. Additionally, these materials exhibit diverse optical and mechanical properties. Silica exhibits a broad range of drawing circumstances as a result of its melting point, which may range from 1400 to 2350 °C. Conversely, silicon and gold exhibit a significant decrease in viscosity if they exceed their melting temperature Tm, indicating a shift in their phase. However, silica can be used as a "cladding" material, where a silicon or gold "core" is put into the preform and then heated above the core's melting temperature (T_m) during the drawing process. Within this particular setting, the cladding functions as a structural framework that limits and controls the movement of the low-viscosity core material.

By examining this example, we can identify the overall limitations of creating pre-forms with multi-materials and the specific requirements for the process of drawing. At least one material, often a glass or polymer, must be amorphous and capable of withstanding devitrification during heat drawing. The "backbone" material serves as an external cladding that provides support to the other materials during the drawing process and ensures the structural integrity of the fiber's cross-section. When choosing amorphous material elements, it is important to favor those with softening temperatures that coincide. For the crystalline constituents, their melting temperature (T_m) should be lower than the drawing temperature. It is imperative that the drawing temperature remains below the boiling point of the primary material. Moreover, it is essential to exercise prudence to avoid the emergence of fluid instabilities that might develop when the viscosity of the substances is reduced and the lateral dimensions are decreased. In order to avoid mechanical fractures resulting from differences in thermal expansion, it is essential for the materials to have thermal expansion coefficients that are relatively similar within the temperature range up to the drawing temperature.

It is worth noting that it is feasible to incorporate more than two elements that satisfy the previously described requirements into the same multi-material fiber. This feature enables the creation of electrical and optoelectronic devices directly within the fiber.

2.3 Multi-material Preform Fabrication

Up till now, four main methods have been used for multi-material preform preparation: (i) the rod-in-tube technique, [36, 37] (ii) extrusion, [38, 39], (iii) The stack-and-draw method [18, 40], and (iv) thin-film-rolling technique [20]. The selection is determined by the specific materials employed and the desired transverse structure. These important methods are illustrated below:

2.3.1 Rod-in-Tube Approach

The process depicted in Fig. 2.2a demonstrates the rod-in-tube method, wherein a preform featuring a core-cladding structure is produced by inserting a rod composed of one material into a tube composed of another material. Alternatively, one can introduce a powdered substance into the tube and then seal the preform, allowing for the inclusion of a diverse array of materials within the central part. If the melting temperature (T_m) of the core rod or powder is lower than the temperature at which the material is drawn, we refer to the process as the molten-core-in-tube technique. The efficacy of this technique for creating fibers with multiple materials may be attributed to the groundbreaking research conducted by E. Snitzer in 1989, [41]. Snitzer employed a combination of selective volatilization and the rod-in-tube approach to fabricate a fiber with a soft-glass core and a silica cladding. In this study, the temperature at which the drawing process took place was greater than the melting temperature (T_m) of a soft-glass core. As a result, the high vapor pressure caused some compounds to evaporate from the core, resulting in a glass core with a distinct residue composition.

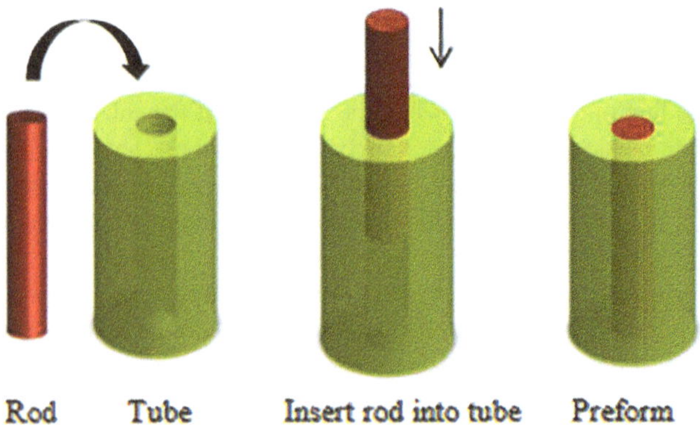

Fig. 2.2a General methodologies for multi-material fiber preform fabrication (Rod-in-tube). (Reprinted from [39])

2.3.2 *Extrusion*

Extrusion is a manufacturing technique that involves forcing a malleable material through a die using pressure, resulting in the creation of objects with consistent cross-sectional profiles. J. Bramah obtained a patent for the initial extrusion method used to manufacture lead pipes. Later, in the 1970s, Roeder expanded this technique to include several types of glass such as soda-lime-silica, lead silicate, calcium aluminate, and boric oxide. Material in the shape of a rod, commonly called a billet, is inserted within a sleeve held in a furnace (a vertically stacked billet is depicted in Fig. 2.2b for demonstration). The billet is heated to the temperature at which the integrated ingredients become soft, and pressure is applied to force the material through a die that gives shape to the extruded preform rod. This rod is then pulled into a fiber. The transverse structure of the extruded rod may be designed by utilizing a multi-material billet, which comprises vertically stacked disks [38, 39].

2.3.3 *Stack-and-Draw Approach*

The stack-and-draw technique has been extensively utilized in the production of preforms for microstructured fibers, photonic crystal fibers (PCFs), and photonic bandgap (PBG) fibers [18]. Bell Labs is credited with first implementing the stack-and-draw approach for producing optical fibers in 1974, as part of their early research on silica fibers. This involved the fabrication of a fiber that possesses a core that is suspended and enclosed by air. A preform (as shown in Fig. 2.2c) can be created by assembling rods, tubes, and/or plates made from one or more materials. The dimensions of the preform are selected based on the desired fiber structure. Iterative stack-and-draw procedures can be employed to achieve the desired dimensions and create intricate transverse formations.

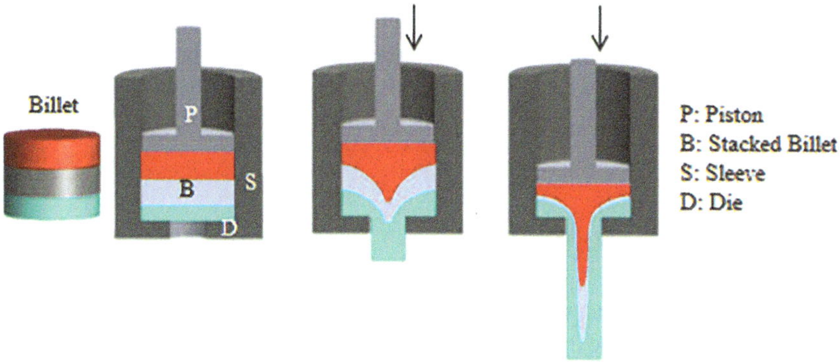

Fig. 2.2b General methodologies for multi-material fiber preform fabrication (Extrusion). (Reprinted from [39])

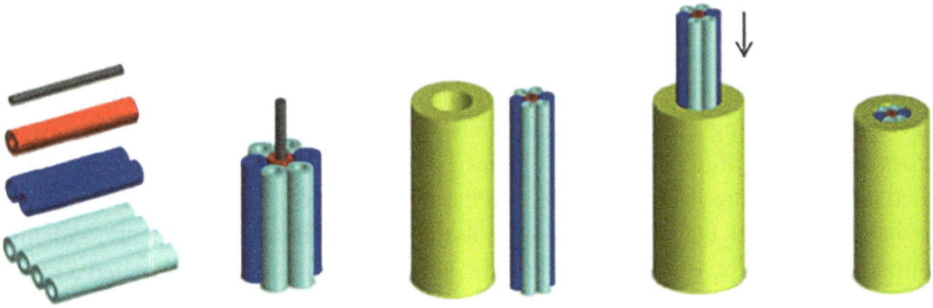

Fig. 2.2c General methodologies for multi-material fiber preform fabrication (Stack and Draw Approach). (Reprinted from [39])

2.3.4 Thin-Film Rolling

Polymers can be incorporated into a fiber using the three ways listed above. A novel technique for incorporating a polymer into a preform entails rolling a thin polymer film and subsequently subjecting it to thermal consolidation in a vacuum at a temperature higher than the glass transition temperature of the constituent components. This process causes the different films to merge together. Figure 2.3 provides a schematic representation of the process, illustrating the manufacturing method used to create a multi-material PBG fiber [42–46]. The thin-film-rolling technique may be repeatedly applied to achieve intricate cross-sections. As an illustration, empty spaces might be included in a building that has undergone consolidation. Conductors can be used to fill these empty spaces and then enclosed within the preform by adding more thin films and consolidating them again. The multistep preform preparation techniques serve as the foundation for several photonic and optoelectronic multi-material-fiber devices [20, 26, 47, 48]. The transmission spectra of fibers with varying periods of the multilayer structure that lines the hollow core demonstrate the precise manipulation of layer thicknesses at the nanoscale. These spectra encompass the whole visible range. Hollow-core multi-material photonic bandgap (PBG) fibers have been previously utilized in many applications.

2.4 Deep-Coating Approach

In the preceding sections, we have outlined different multi-material fibers that have been created utilizing the overall approach of preform-to-fiber manufacturing. In this section, we quickly describe some intriguing exceptions. A technique proposed involves using a long steel wire that is dipped in organic solutions in a series of phases [49]. The outcome is a fiber that possesses a multi-layer covering across its

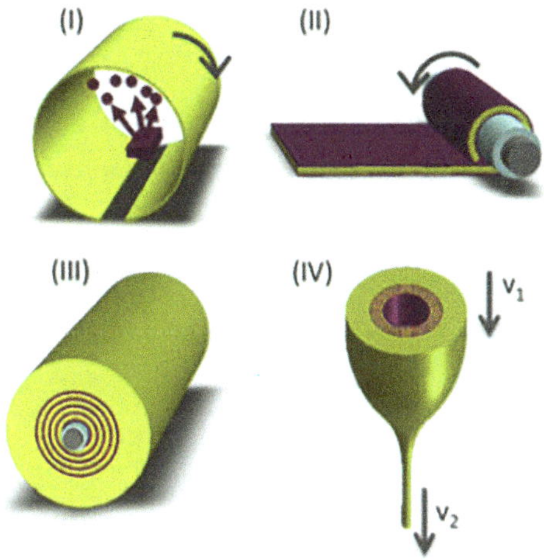

Fig. 2.3 Fabrication and characterization of PBG fibers. The process of manufacturing PBG fiber. (**I**)The chalcogenide glass is deposited onto both sides of the polymer sheet using thermal evaporation. (**II**) The multilayer film is then wrapped around a mandrel coated with Teflon, and more layers of polymer cladding are added for mechanical reinforcement. (**III**) The complete construction is subjected to thermal consolidation in a vacuum environment, causing the materials to merge and form a single, solid preform. (**IV**) The initial shape is then transformed into a long fiber by subjecting it to unidirectional tension and thermal drawing, resulting in hundreds of meters of fiber. The relationship between the down-feed speed (v_1) of the preform and the draw speed (v_2) of the fiber determines the thickness of the final layer. (Reprinted from [39])

whole length, resulting in the creation of an organic photovoltaic cell with an approximate efficiency of 3%. The second approach entails employing a silica fiber that has voids (either in the form of a hollow-core fiber or a photonic crystal fiber) as a framework for depositing conventional crystalline semiconductors via vapor deposition. The high-pressure microfluidic chemical deposition (HPMCD) approach enables the integration of a broader selection of materials into the fiber, including single-crystal semiconductors and polycrystalline elemental or compound semiconductors. In order to surmount constraints in mass transport, silica microstructured optical fibers (MOFs) utilize high-pressure flow, which can vary between 2 and 1000 MPa. This process leads to the even, compact, and conformal deposition of material onto the inner walls of the hollow pores. The resultant homogeneity is on the scale of 0.1 nm. This methodology combines two established manufacturing techniques, namely chemical vapor deposition (CVD) and silica fiber drawing. The fundamental concept is illustrated in Fig. 2.4a, b, whereby tiny amounts of a high-pressure precursor are heated within the interior of an MOF. This precursor decomposes during heating and deposits on the walls in an amorphous state to ensure a smooth bonding. Subsequently, annealing is performed to induce crystallization. Figure 2.4b illustrates the presence of doped semiconductor layers and metals that have been placed or laid down [50]. As the thickness of the annular deposited layer increases, the size of the central pores through which

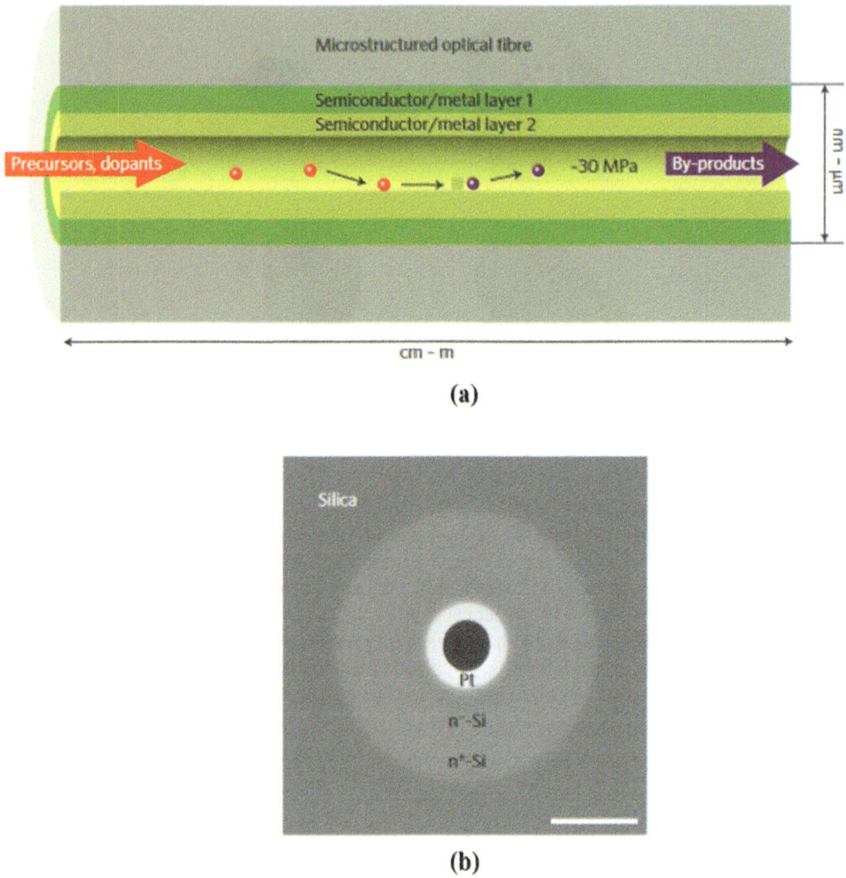

Fig. 2.4 Integration of semiconductor junctions in MOFs. (**a**) Illustration of HPCVD in the MOF pores. (**b**) A Pt/n-Si Schottky junction formed by sequential deposition of phosphorous doped n + -Si, n-Si, and platinum layers (Scale bar: 5 μm) [50]. (Reproduced with permission from [50])

depositions take place decreases until they are totally blocked and the flow stops. The vacant cavities within the MOFs are regarded as micro- or nanoscale compartments for chemical reactions. A device made of fibers, created using this method, allowed for the regulation of 1.55-lm directed light via free-carrier absorption, which was facilitated by a 532-nm pump pulse [51]. Additional instances encompass the creation of a ZnSe-core fiber [52], the use of in-fiber Si and Ge wires and tubes as field effect transistors [53], and the development of in-fiber crystalline Si p–n homojunction and Pt/n- type Si-heterojunctions [50]. The HPCVD process is versatile and may be used to fill various capillary core dimensions, as well as a multitude of micro- and nanoscale holes in MOFs. An inherent limitation of this approach is the restricted lengths of fiber devices manufactured, in contrast to those obtained using fiber drawing.

2.5 Hollow-Core Multi-material PBG Fibers

This section presents a comprehensive review of multi-material fibers, which are utilized for the conventional purpose of guiding light along the fiber's axis, as well as for innovative applications including transversely active photonic fiber devices. The team headed by Y. Fink at MIT has effectively demonstrated one of the early examples of multi-material optical fibers [54]. We see photonic bandgap (PBG) fibers that possess an all-solid cladding capable of confining light within a hollow core. This is achieved by utilizing two materials with a substantial disparity in refractive index. Furthermore, we shall now proceed to elucidate additional examples of photonic multi-material fibers.

A flat omnidirectional mirror may be created by utilizing a one-dimensional (1D) periodic stack of alternating high- and low-index materials that have a significant difference in refractive index. This mirror is capable of reflecting light that comes from any angle and polarization within a certain range of wavelengths. The mirror may be applied to the inner surface of a hollow-core fiber, allowing for the directing of light along the fiber's air core in an axial direction [54]. Although there were early indications of a waveguiding mechanism similar to this since the 1970s, it was not until 2002 that the first multi-material hollow-core fiber was successfully produced using the usual preform-to-fiber drawing technique. The difficulty is twofold. Initially, it is necessary to choose a set of materials that possess both suitable thermomechanical characteristics and a significant disparity in refractive index. Furthermore, in order to manufacture low-loss fibers, it is imperative to uphold the multilayer structure with low-scattering interfaces at both the micro- and nanoscales. Currently, a number of material combinations have been discovered that provide optical transmission fibers with little loss when fabricated correctly. A common pair comprises a low-refractive-index polymer and a high-refractive-index chalcogenide glass, both of which are amorphous and thermomechanically compatible for fiber drawing [43, 44]. The procedure for manufacturing hollow-core multi-material fibers with photonic band gaps (PBGs) that cover the infrared to ultraviolet range is described in Fig. 2.3 (I–IV). The precise manipulation of layer thicknesses at the nanoscale scale is showcased by the transmission spectra, which cover the whole visible range, of fibers that vary solely in the period of the multilayer structure that lines the hollow core. These hollow-core multi-material PBG fibers have been used in several applications thus far.

2.6 Future Prospects

We want to provide a comprehensive summary of the diverse array of innovative concepts now being explored in the emerging domain of multi-material fibers. The area has experienced significant growth over the past decade, and it is unfeasible to encompass all the topics being investigated in this concise overview. An example of a missing aspect is the new research on creating metamaterials using multi-material fibers [55–59]. Metamaterials are man-made structures that manipulate light and allow for unusual optical effects, such as cloaking [37]. The manipulation of THz radiation transmitted via the transverse cross-sections of a polymer fiber has been demonstrated by creating arrays of metallic objects, allowing for control over the electrical and magnetic resonances. To apply this method to optical wavelengths, it is necessary to decrease the diameters of the features in the multi-material fibers even more, while also preventing any instabilities caused by fluids inside the fibers. Other examples include the use of gold in silica fibers for plasmonic studies and the use of soft glasses within silica fibers for nonlinear applications in the mid-infrared region [60–66]. The optical fibers used are made from sapphire and have been specifically designed to exhibit Brillouin scattering features. These fibers are well-suited for applications such as sensors and high-energy fiber lasers.

Additional investigation is required to enhance the compactness of optical, electrical, and optoelectronic components integrated into a multi-material fiber. The development of fibers having these functions, which can be directly woven into fabrics, would be a significant achievement in this sector, combining optics, electronics, and textiles. The utilization of in-fiber capillary instabilities for the creation of three-dimensional macroscopic synthetic photonic structures, with precise control over the distribution of materials at the nanoscale, would allow for unparalleled manipulation of the properties of light and sound. Undoubtedly, there are more unexpected events to come. The advancement in multi-material fibers discussed in this chapter indicates a revival in fiber manufacturing that is expected to last for a considerable duration.

References

1. G.P. Agrawal, *Fiber-Optic Communication Systems*, 4th edn. (Wiley, Hoboken, 2010)
2. R. Ramaswami, K.N. Sivarajan, G.H. Sasaki, *Optical Networks: A Practical Perspective*, 3rd edn. (Morgan Kaufmann, San Francisco, 2009)
3. R.K. Chourasia, S. Prasad, V. Singh, Voltage-tunable pass band in a cylindrical multilayered structure containing PMMA and 0.67PMN–0.33PT single crystal as a defect layer. Opt. Quant. Electron. **48**, 539 (2016)
4. N.K. Chourasia, A. Srivastava, V. Kumar, R.K. Chourasia, Doubly electrically tuned cylindrical Bragg fiber waveguide inline optical filter for multiwavelength LASER applications. Mater. Today Commun. **25**, 101620 (2020)
5. A.G. Mignani, F. Baldini, Biomedical sensors using optical fibres. Rep. Prog. Phys. **59**, 1–28 (1996)
6. Z. Zhou, J. He, M. Huang, J. He, G. Chen, Casing pipe damage detection with optical fiber sensors: A case study in oil well constructions. Adv. Civil Eng. **9**, 638967 (2010)
7. N. Mohamed, I. Jawhar, J. Al-Jaroodi, L. Zhang, Sensor network architectures for monitoring underwater pipelines. Sensors **11**(11), 10738–10764 (2011)
8. R.K. Chourasia, C.S. Yadav, A. Upadhyay, N.K. Chourasia, V. Singh, Analysis of Bragg fiber waveguide having a defect layer for biosensing application. Optik **200**, 163400 (2020)
9. R.K. Chourasia, S. Prasad, V. Singh, Performance study of a liquid-core Bragg fiber sensor in the presence of a defect layer. Opto-Electron. Rev. **25**, 215–221 (2017)

10. N.K. Chourasia, N. Bihari, R.K. Chourasia, An optofluidic Bragg fiber sensor for estimating adulterants in a temperature-dependent molar fraction of hydrated mono-alcohol fuels. Heliyon **8**(9), e10532 (2022)
11. R.K. Chourasia, C.S. Yadav, A. Upadhyay, N.K. Chourasia, V. Singh, Comparative study of defect mode intensity and wavelength modulation in Bragg fiber waveguide sensors. Optik **204**, 164198 (2020)
12. N.K. Chourasia, A. Srivastava, V. Kumar, R.K. Chourasia, Comparative performance study of liquid core cylindrical Bragg fibre waveguide biosensors. Pramana J. Phys. **95**, 9 (2021)
13. A. Shadab, S.K. Raghuwanshi, Development and sensitivity analysis of rGO-TiO_2 coated eFBG sensor for the detection of ethanol in petrochemicals. IEEE Sensors J. **22**(13), 12913–12920 (2022). https://doi.org/10.1109/JSEN.2022.3180205
14. N.K. Chourasia, A. Srivastava, V. Kumar, R.K. Chourasia, Optimizing temperature-dependent molar volume fraction of biodiesel fuel in a pseudo-binary mixture through Bragg fiber waveguide sensor having defect layer. Fuel **293**, 120489 (2021)
15. R.K. Chourasia, V. Singh, *Cylindrical Bragg waveguide-based temperature sensors*, in IEEE Uttar Pradesh Section International Conference on Electrical, Computer and Electronics Engineering (UPCON) 87-90 (2016)
16. R.K. Chourasia, S. Prasad, V. Singh, A novel Bragg fiber waveguide-based narrowband inline optical filter. Optik **169**, 269–276 (2018)
17. T. Webber, *Advances in Fiber Lasers for the Materials Processing Market*, in Quantum Electronics and Laser Science Conference (QELS) on Advances in High-Power Lasers and their Applications III: Processing (JTh4I), San Jose, May 6 (2012)
18. P. Russell, Photonic crystal fibers. Science **299**, 358–362 (2003)
19. J.D. Joannopoulos, S.G. Johnson, J.N. Winn, R.D. Meade, P. Crystals, *Molding the Flow of Light*, 2nd edn. (Princeton University Press, Princeton, 2008)
20. A.F. Abouraddy, M. Bayindir, G. Benoit, et al., Towards multimaterial multifunctional fibres that see, hear, sense and communicate. Nat. Mater. **6**, 336–347 (2007)
21. M. Bayındir, F. Sorin, S. Hart, O. Shapira, J.D. Joannopoulos, Y. Fink, Metal-insulator-semiconductor optoelectronic fibres. Nature **431**, 826–829 (2004)
22. F. Sorin et al., Multimaterial photodetecting fibers: A geometric and structural study. Adv. Mater. **19**, 3872–3877 (2007)
23. A.F. Abouraddy et al., Large-scale optical-field measurements with geometric fibre constructs. Nat. Mater. **5**, 532–536 (2006)
24. M. Bayindir, A.F. Abouraddy, J. Arnold, J.D. Joannopoulos, Y. Fink, Thermal-sensing fiber devices by multimaterial codrawing. Adv. Mater. **18**, 845–849 (2006)
25. M. Bayindir et al., Integrated fibers for self monitored optical transport. Nat. Mater. **4**, 820–824 (2005)
26. F. Sorin et al., Exploiting collective effects of multiple optoelectronic devices integrated in a single fiber. Nano Lett. **9**, 2630–2635 (2009)
27. S. Egusa et al., Multimaterial piezoelectric fibres. Nat. Mater. **9**, 643–648 (2010)
28. N. Chocat, G. Lestoquoy, Z. Wang, D.M. Rodgers, J.D. Joannopoulos, Y. Fink, Piezoelectric fibers for conformal acoustics. Adv. Mater. **39**, 5327–5332 (2012)
29. D.S. Deng, N.D. Orf, S. Danto, A.F. Abouraddy, J.D. Joannopoulos, Y. Fink, Processing and properties of centimeter-long, in-fiber, crystalline-selenium filaments. Appl. Phys. Lett. **96**, 023102 (2010)
30. M. Yaman et al., Arrays of indefinitely long uniform nanowires and nanotubes. Nat. Mater. **10**, 494–501 (2011)
31. J.J. Kaufman, G. Tao, S. Shabahang, D.S. Deng, Y. Fink, A.F. Abouraddy, Thermal drawing of high-density macroscopic arrays of well-ordered sub-5-nm-diameter nanowires. Nano Lett. **11**, 4768–4773 (2011)
32. S. Shabahang, J.J. Kaufman, D.S. Deng, A.F. Abouraddy, Observation of the plateau-Rayleigh capillary instability in multi-material optical fibers. Appl. Phys. Lett. **99**, 161909 (2011)

33. J.J. Kaufman et al., Structured spheres generated by an in-fibre fluid instability. Nature **487**, 463–467 (2012)
34. T. Li (ed.), *Optical Fiber Communications: Fiber Fabrication* (Academic Press, Waltham, 1985)
35. H. Tokiwa, Y. Mimura, T. Nakai, O. Shinbori, Fabrication of long single-mode and multimode fluoride glass fibres by the double-crucible technique. Electron. Lett. **21**(24), 1131–1132 (1985)
36. B.L. Scott, K. Wang, G. Pickrell, Fabrication of n-type silicon optical fiber. IEEE Photon. Technol. Lett. **21**(24), 1798–1800 (2009)
37. H.K. Tyagi et al., Plasmon resonances on gold nanowires directly drawn in a step-index fiber. Opt. Lett. **35**(15), 2573–2575 (2010)
38. D.J. Gibson, J.A. Harrington, Extrusion of hollow waveguide preforms with a one-dimensional photonic bandgap structure. J. Appl. Phys. **95**(8), 3895–3900 (2004)
39. G. Tao, A.M. Stolyarov, A.F. Abouraddy, Multimaterial fibers. Int. J. Appl. Glass. Sci. **3**(4), 349–368 (2012)
40. M. Liao et al., Fabrication and characterization of a chalcogenide-tellurite composite microstructure fiber with high nonlinearity. Opt. Express **17**(24), 21608–21614 (2009)
41. E. Snitzer, R. Tumminelli, SiO2-clad fibers with selectively volatilized soft-glass cores. Opt. Lett. **14**(14), 757–759 (1989)
42. B. Temelkuran, S.D. Hart, G. Benoit, J.D. Joannopoulos, Y. Fink, Wavelength-scalable hollow optical fibres with large photonic band- gaps for CO2 laser transmission. Nature **420**, 650–653 (2002)
43. K. Kuriki et al., Hollow multilayer photonic bandgap fibers for NIR applications. Opt. Express **12**(8), 1510–1517 (2004)
44. Z. Ruff, D. Shemuly, X. Peng, O. Shapira, Z. Wang, Y. Fink, Polymer-composite fibers for transmitting high peak power pulses at 1.55 microns. Opt. Express **18**(15), 15697–15703 (2010)
45. D. Shemuly, A.M. Stolyarov, Z.M. Ruff, L. Wei, Y. Fink, O. Shapira, Preparation and transmission of low-loss azimuthally polarized pure single mode in multimode photonic band gap fibers. Opt. Express **20**(6), 6029–6035 (2012)
46. A.M. Stolyarov et al., Enhanced chemiluminescent detection scheme for trace vapor sensing in pneumatically-tuned hollow core photonic bandgap fibers. Opt. Express **20**(11), 12407–12415 (2012)
47. A.M. Stolyarov, L. Wei, F. Sorin, G. Lestoquoy, J.D. Joannopoulos, Y. Fink, Fabrication and characterization of fibers with built-in liquid crystal channels and electrodes for transverse incident-light modulation. Appl. Phys. Lett. **101**, 011108 (2012)
48. A.M. Stolyarov et al., Microfluidic directional emission control of an azimuthally polarized radial fibre Laser. Nat. Photonics **4**, 229–233 (2012)
49. M.R. Lee, R.D. Eckert, K. Forberich, G. Dennler, C.J. Brabec, R.A. Gaudiana, Solar power wires based on organic photovoltaic materials. Science **324**, 232–235 (2009)
50. R. He et al., Integration of gigahertz-bandwidth semiconductor devices inside microstructured optical fibres. Nat. Photonics **6**, 174–179 (2012)
51. D.-J. Won et al., All-optical modulation of laser light in amorphous silicon-filled microstructured optical fibers. Appl. Phys. Lett. **91**, 161112 (2007)
52. J.R. Sparks et al., Zinc selenide optical fibers. Adv. Mater. **23**, 1647–1651 (2011)
53. C.E. Finlayson, A. Amezcua-Correa, P.J.A. Sazio, N.F. Baril, J.V. Badding, Electrical and Raman characterization of silicon and germanium-filled microstructed optical fibers. Appl. Phys. Lett. **90**, 132110 (2007)
54. Y. Fink, D.J. Ripin, S. Fan, C. Chen, J.D. Joannopoulos, E.L. Thomas, Guiding optical light in air using an all-dielectric structure. J. Lightwave Technol. **17**(11), 2039–2041 (1999)
55. A. Wang et al., Fiber metamaterials with negative magnetic permeability in the terahertz. Opt. Mater. Express **1**(1), 115–120 (2011)
56. A. Tuniz et al., Stacked-and-drawn metamaterials with magnetic resonances in the terahertz range. Opt. Express **19**(17), 16480–16490 (2011)

57. A. Tuniz et al., Spatial dispersion in three-dimensional drawn magnetic metamaterials. Opt. Express **20**(11), 11924–11935 (2012)
58. R.C. McPhedran, I.V. Shadrivov, B.T. Kuhlmey, Y.S. Kivsha, Metamaterials and metaoptics. NPG Asia Mater. **3**, 100–108 (2011)
59. N. Granzow, S.P. Stark, M.A. Schmidt, A.S. Tverjanovich, L. Wondraczek, P.S.J. Russell, Supercontinuum generation in chalcogenide-silica step-index fibers. Opt. Express **19**(21), 21003–21010 (2011)
60. J. Hou, D. Bird, A. George, S. Maier, B.T. Kuhlmey, J.C. Knight, Metallic mode confinement in microstructured fibres. Opt. Express **16**(9), 5983–5990 (2008)
61. C.G. Poulton, M.A. Schmidt, G.J. Pearce, G. Kakarantzas, P.S.J. Russell, Numerical study of guided modes in arrays of metallic nanowires. Opt. Lett. **32**(12), 1647–1649 (2007)
62. M.A. Schmidt, L.N.P. Sempere, H.K. Tyagi, C.G. Poulton, P.S.J. Russell, Waveguiding and Plasmon resonances in two-dimensional photonic lattices of gold and silver nanowires. Phys. Rev. B **77**, 033417 (2008)
63. H.W. Lee, M.A. Schmidt, H.K. Tyagi, L.N. Prill Sempere, P.S.J. Russell, Polarization-dependent coupling to Plasmon modes on sub- micron gold wire in photonic crystal fiber. Appl. Phys. Lett. **93**(11), 111102 (2008)
64. H.W. Lee et al., Pressure-assisted melt-filling and optical characterization of Au nano-wires in microstructured fibers. Opt. Express **19**(13), 12180–12189 (2011)
65. N. Granzow, P. Uebel, M.A. Schmidt, A.S. Tverjanovich, L. Wondraczek, P.S.J. Russell, Bandgap guidance in hybrid chalcogenide–silica photonic crystal fibers. Opt. Lett. **36**(13), 2432–2434 (2011)
66. P. Dragic, T. Hawkins, P. Foy, S. Morris, J. Ballato, Sapphire- derived all-glass optical fibres. Nat. Photonics **6**, 627–633 (2012)

Chapter 3
Various Analytical Techniques, Theorems, and Formalisms

Abstract In the first chapter of this book, we explain why we chose Bragg fiber for our studies and developments in the past. These biologically inspired multilayer hollow fibers are found to have a structure that is very similar to that of naturally occurring fruit. The various fabrication techniques available for the experimental construction of such a fiber structure are introduced to the reader in the second chapter. The present chapter continues the investigation by applying theoretical modeling and mathematical analysis to the Bragg fiber structure. It does so by discussing in detail the various analytical methods accessible, as well as relevant theorems and formalism. A comparison of these various approaches to analysis is finally presented.

3.1 Asymptotic Technique

Bragg fibers (BFs) are optical waveguides that employ cylindrical Bragg reflection instead of total internal reflection to accomplish light confinement. The fibers exhibit a distinctive and compelling property, specifically the capability to direct light within an air core. This feature has received considerable interest in the scientific world, as seen by the vast research undertaken on the topic [1–7]. Air core fibers have numerous potential benefits in terms of optical transmission. An advantage of this is the capacity to attain reduced absorption loss, hence greatly improving the efficiency of signal propagation. Moreover, air core fibers possess a greater threshold power for nonlinear effects, which might be advantageous in situations where it is necessary to minimize or regulate nonlinear events. Previous studies have emphasized these benefits [8]. Researchers completed a study that provided an experimental demonstration of waveguiding in Bragg fiber. Theoretical analysis of Bragg fibers is considerably more complex in comparison to that of planar waveguides [1, 2]. The major reason is that the Bloch theorem does not hold true for cylindrically symmetric geometries, such as Bragg fibers. Therefore, identifying the restricted modes presents a difficult undertaking in terms of discovering a suitable equation. This chapter will concentrate on the examination of guided modes in Bragg fiber utilizing an asymptotic methodology within the asymptotic limit. The application of the

R. K. Chourasia, A. Katti, *Bragg Fibers*,
https://doi.org/10.1007/978-3-031-65164-9_3

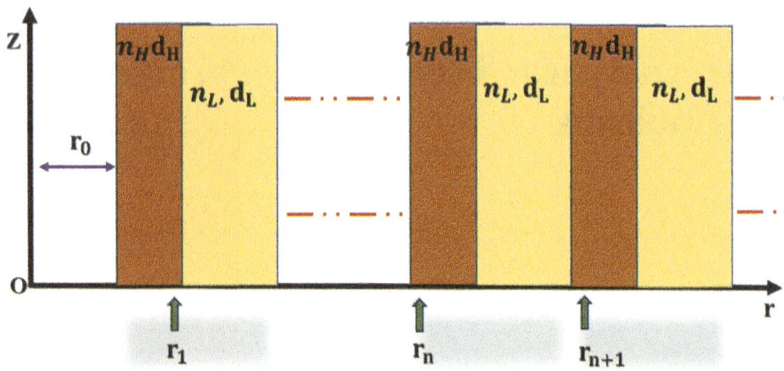

Fig. 3.1 The diagram depicts a schematic illustration of the proposed Bragg fiber's *r-z* cross section. The fiber core possesses a refractive index denoted as n_c and is characterized by a radius r_1. The composition of the fiber cladding comprises alternating layers of materials with high and low refractive indices, organized in pairs. The layer with a high refractive index is denoted by n_H and has a thickness of l_H. The layer possessing a lower refractive index is distinguished by a refractive index value designated as n_L and a thickness designated as l_L

asymptotic technique has been found to greatly simplify the problem and provide a very accurate estimate of the exact answer, even in circumstances when the radius of the hollow core is rather small.

This research focuses on a Bragg fiber structure of a hollow core or a core having a low refractive index (n_C). The core is surrounded by alternating layers of high refractive index (n_H) and low refractive index (n_L). The remaining physical properties of the Bragg fiber are indicated in Fig. 3.1 below. In the context of guided modes, it is possible to neglect the temporal and z-dependency, which may be represented as exp[i(βz-ωt)], as well as the azimuthal dependence, which can be expressed as cos($l\phi$) [9–12]. When utilizing the asymptotic formulae for the Bessel function as $kr \rightarrow \infty$, the fluctuations of E_z and H_z in the radial direction are described

$$E_z = a_c J_l(k_c r); (0 < r < r_1) \tag{3.1a}$$

$$E_z = \frac{a_n \exp\left[ik_H(r - r_n)\right] + b_n \exp\left[ik_H(r - r_n)\right]}{\sqrt{k_H r}}; (r_n < r < r_n + l_H) \tag{3.1b}$$

$$E_z = \frac{a'_n \exp\left[ik_L(r - r'_n)\right] + b'_n \exp\left[ik_L(r - r'_n)\right]}{\sqrt{k_L r}}; (r'_n < r < r'_n + l_L) \tag{3.1c}$$

$$H_z = c_c J_l(k_c r); (0 < r < r_1) \tag{3.1d}$$

$$H_z = \frac{c_n \exp\left[ik_H(r - r_n)\right] + d_n \exp\left[ik_H(r - r_n)\right]}{\sqrt{k_H r}}; (r_n < r < r_n + l_H) \tag{3.1e}$$

$$H_z = \frac{c'_n \exp\left[ik_L(r - r'_n)\right] + d'_n \exp\left[ik_L(r - r'_n)\right]}{\sqrt{k_L r}}; \left(r'_n < r < r'_n + l_L\right) \tag{3.1f}$$

where $k_c = \left[n_c^2\left(\frac{\omega}{c}\right)^2 - \beta^2\right]^{\frac{1}{2}}$; $k_H = \left[n_H^2\left(\frac{\omega}{c}\right)^2 - \beta^2\right]^{\frac{1}{2}}$ and $k_L = \left[n_L^2\left(\frac{\omega}{c}\right)^2 - \beta^2\right]^{\frac{1}{2}}$

By having knowledge of E_z and H_z, it becomes straightforward to determine the remaining field components by the utilization of their respective derivatives [10]. The continuity of E_z, E_ϕ, H_z, and H_ϕ at the contact between two adjacent layers is a requirement imposed by the boundary conditions. By considering the asymptotic limit, we may determine the matrix relations for $E\phi$ and $H\phi$ by retaining only the leading terms:

$$\begin{pmatrix} a_{n+1} \\ b_{n+1} \end{pmatrix} = \begin{bmatrix} A_{TM} & B_{TM} \\ B_{TM}^* & A_{TM}^* \end{bmatrix} \begin{pmatrix} a_n \\ b_n \end{pmatrix} \tag{3.2a}$$

$$\begin{pmatrix} c_{n+1} \\ d_{n+1} \end{pmatrix} = \begin{bmatrix} A_{TE} & B_{TE} \\ B_{TE}^* & A_{TE}^* \end{bmatrix} \begin{pmatrix} c_n \\ d_n \end{pmatrix} \tag{3.2b}$$

where A_{TE}, B_{TE}, A_{TM} and B_{TM} are defined as:

$$A_{TE} = \exp(ik_H l_H)\left[i\frac{k_H^2 + K_L^2}{2k_H k_L}\sin(k_L l_L) + \cos(k_L l_L)\right] \tag{3.3a}$$

$$B_{TE} = i\exp(-ik_H l_H)\frac{k_H^2 - K_L^2}{2k_H k_L}\sin(k_L l_L) \tag{3.3b}$$

$$A_{TM} = \exp(ik_H l_H)\left[i\frac{n_2^4 k_H^2 + n_1^4 K_L^2}{2k_H k_L}\sin(k_L l_L) + \cos(k_L l_L)\right] \tag{3.3c}$$

$$B_{TM} = i\exp(-ik_H l_H)\frac{n_2^4 k_H^2 - n_1^4 K_L^2}{2k_H k_L}\sin(k_L l_L) \tag{3.3d}$$

Since A_{TE}, B_{TE}, A_{TM}, and B_{TM} are the same for all cladding layers, therefore the Bloch theorem can be applied to the cladding fields:

$$(a_{n+1}\ b_{n+1}) = \exp(iK_{TM}\wedge)(a_n\ b_n) \tag{3.4a}$$

$$(c_{n+1}\ d_{n+1}) = \exp(iK_{TE}\wedge)(c_n\ d_n) \tag{3.4b}$$

$$\exp(iK_{TM}\wedge) = \mathrm{Re}(A_{TM}) \pm \left\{[\mathrm{Re}(A_{TM})]^2 - 1\right\}^{\frac{1}{2}} \tag{3.4c}$$

$$\exp(iK_{TE}\wedge) = \mathrm{Re}(A_{TE}) \pm \left\{[\mathrm{Re}(A_{TE})]^2 - 1\right\}^{\frac{1}{2}} \tag{3.4d}$$

The results unambiguously demonstrate that in the asymptotic limit, the characteristics of Bragg fiber cladding are very similar to those of planar Bragg stacks. As a result, numerous findings for planar Bragg stacks might be readily utilized for Bragg fibers. Here, $\mathrm{Re}(A_{TE}) > 1$ denotes the TE Bragg reflection gap. Quarter wave stack condition should be utilized to attain optimal confinement in Bragg fibers.: $k_H l_H = k_L l_L = \frac{\pi}{2}$. In order to solve for the propagation constant β, we need to match the values of E_z, E_ϕ, H_z and H_ϕ at $(r = r_1)$, which gives us:

$$\frac{\omega^2}{c^2} n_c^2 \left[\frac{J_l'(k_c r_1)}{J_l(k_c r_1)} + i \frac{k_c n_H^2 \exp\left(iK_{TM} \wedge\right) - A_{TM} - B_{TM}}{k_H n_C^2 \exp\left(iK_{TM} \wedge\right) - A_{TM} + B_{TM}} \right] \times \left[\frac{J_l'(k_c r_1)}{J_l(k_c r_1)} + i \frac{k_c}{k_1} \frac{\exp\left(iK_{TE} \wedge\right) - A_{TE} - B_{TE}}{\exp\left(iK_{TE} \wedge\right) - A_{TE} + B_{TE}} \right] = \frac{\beta^2 l^2}{k_c^2 r_1^2} \tag{3.5}$$

It is evident that the guided modes consist of a combination of TE and TM modes for any value of $l \neq 0$. In the singular circumstance where l equals zero, only then can the guided modes be categorized as TE or TM modes.

Equation 3.5 governs the general mode dispersion in any Bragg fiber. It may seem that asymptotic approximation is only valid for a core radius of at least several wavelengths. The modes in 1D planar stacks are quite similar to TE/TM and other low-order azimuthal hybrid modes [13–17]. With their asymptotic formulations, the asymptotic method provides an approximation for the Bessel functions which characterize the cladding layers [17]. The planned Bragg stacks are a good approximation for the periodic cylindrical claddings. The suggested structure loses its identity in this technique, which is able to considerably cut computing time but is also ineffective at interacting with and solving layer-by-layer designs since only the core and the first few layers are examined analytically. While using air-core Bragg fibers, the benefit of this approach is that it offers an analytical Eigen equation for the propagation constant.

3.2 Galerkin Technique

Galerkin's method is a potent technique that provides a suitable solution to the scalar wave equation for optical waveguides that are of practical significance. The effectiveness of this strategy relies on a judicious selection of the orthogonal basis functions. The Galerkin approach employs an orthonormal collection of functions to transform a differential equation into a system of simultaneous linear equations. In [18], Laguerre-Gauss basis functions were employed to conduct a numerical analysis of circular fibers, taking advantage of their suitability for circular geometries. Henry and Verbeek [19] employed trigonometric basis functions and the Galerkin method to determine the scalar electromagnetic field for a two-dimensional optical waveguide with a cross-section of any shape. These basis functions need enclosing the waveguide within an artificial rectangle of sufficient size to ensure that the guided mode's field is insignificantly small at the boundary. Regrettably, determining the appropriate dimensions for the artificial rectangle poses a challenge, as the outcomes are highly contingent on its size. This book focuses on working in a rectangular coordinate and utilizing the Hermite-Gauss basis functions. By naturally fulfilling the boundary criteria at infinity, these functions obviate the necessity of the enclosed rectangle. Therefore, we have a significant degree of freedom in selecting the basis functions required by Galerkin's technique. However, making a wise choice will result in computational benefits. Utilizing the Hermite-Gauss basis functions has several benefits. Initially, a comprehensive and mutually perpendicular collection is established, which fulfills the boundary constraints both at an infinite distance and at the starting point. Furthermore, they accurately depict the modal eigenfunctions for the infinitely stretched parabolic profile in a rectangular waveguide. Ultimately, they result in analytical matrix elements for a broad range of refractive index profiles.

Let's start by examining the scalar wave equation for a waveguide in the rectangular coordinate system:

$$\frac{d^2\zeta}{dx^2}+\frac{d^2\zeta}{dy^2}+k^2\left(n^2(x,y)-n_e^2\right)\zeta=0 \tag{3.6}$$

Where the scalar wave function is represented by ζ, n_e denotes the effective index and k represents the free space wavenumber. Let us consider an untruncated parabolic R.I. profile, for which;

$$n^2(x,y)=n_1^2\left[1-2\Delta\left(\frac{x}{a}\right)^2-2\Delta\left(\frac{y}{b}\right)^2\right] \tag{3.7}$$

where n_1 is the R.I. on the waveguide axis, Δ is the contrast, and "a" and "b" are scaling distances along the "x" and "y" coordinates. Here $(a = b)$ is the case of circular geometry. For the untruncated profile (3.6) and (3.7) hold for all x and all y. The exact solutions to 3.6 in this case are:

$$\varsigma_{i,j}=A_{i,j}H_i\left(\sqrt{V_a}\frac{x}{a}\right)H_j\left(\sqrt{V_b}\frac{x}{b}\right)\exp\left[-V_a\left(\frac{x}{a}\right)^2-V_b\left(\frac{y}{b}\right)^2\right] \tag{3.8}$$

with the corresponding eigenvalues:

$$(n_e)_{i,j}^2=n_1^2\left[1-\frac{(2i+1)\sqrt{2\Delta}}{akn_1}-\frac{(2j+1)\sqrt{2\Delta}}{bkn_1}\right] \tag{3.9}$$

where

$$V_a=ka\sqrt{n_1^2-n_2^2}=kan_1\sqrt{2\Delta} \tag{3.10}$$

$$V_b=kb\sqrt{n_1^2-n_2^2}=kbn_1\sqrt{2\Delta} \tag{3.11}$$

H_i and H_j are Hermite polynomials of order i, j, and $A_{i\cdot j}$ is constant, n_2 is the R.I. at $x = 0$, $y =$ b or $x = a$, $y = 0$. Equations 3.7 and 3.8 are applicable to a circular geometry. The subscripts i and j are used to denote the mode index. The zeroth order term refers to the Gaussian function, which serves as a reliable approximation for the fundamental mode. The selection of values for "a" and "b" in Eq. 3.8 reveals the commonly encountered geometries in actual applications, such as circular, elliptical, square, or rectangular waveguides. However, it is subjective in more intricate geometries.

Furthermore, the Galerkin method has mostly been employed to calculate LP modes in conventional circular fibers with relatively small fluctuations in refractive index. The method utilizes a collection of orthogonal Laguerre-Gauss functions that are coupled to approximate the propagation characteristics of guided modes [18, 20–22]. To address the issue of solving transverse fields in fibers with significant differences in indices in two-dimensional Cartesian coordinates, researchers employed vectorial Galerkin methods utilizing either double series sine functions [23, 24] or Hermite-Gauss functions [25]. However, these methods failed to exploit the circular symmetry of the fibers and proved to be inefficient for layer-by-layer problems.

3.3 Transfer Matrix Technique

The transfer-matrix methodology is a widely recognized theoretical approach in optics and acoustics that is used to study the propagation of electromagnetic or acoustic waves through a layered optical medium. The design of anti-reflective coatings and dielectric mirrors could have significant implications, as demonstrated in previous studies [26–28]. The renowned Fresnel equations elucidate the phenomenon of light reflection at a single interface between two different media [26]. The reflections themselves are partly reflected and then transmitted when there are several contacts, however. These reflections may interact destructively or constructively, depending on the precise optical path length. An unlimited number of reflections across the layered structure add up to the overall reflection of a multilayer optical system. Maxwell's equations dictate that the electric field and magnetic field must satisfy clear continuity conditions at the interfaces of the multilayer structure. This is the foundation of the transfer-matrix approach. If the field solution for the first layer is known at the outset, then the field solution for the multilayered medium may be obtained via a straightforward matrix operation. Subsequently, a stack of layers can be denoted as a system matrix, which is the aggregate of the matrices for each individual layer [29–35]. The final part of the procedure is converting the system matrix into reflection and transmission coefficients.

This section provides a comprehensive overview of the many steps and methodologies employed to investigate the optical properties of the Bragg fiber structure. Figure 3.2 depicts the cross-sectional view and refractive index profile of the suggested Bragg fiber. The core of the chosen Bragg fiber is surrounded by alternating layers of dielectric materials with high and low refractive indices, creating a periodic structure.

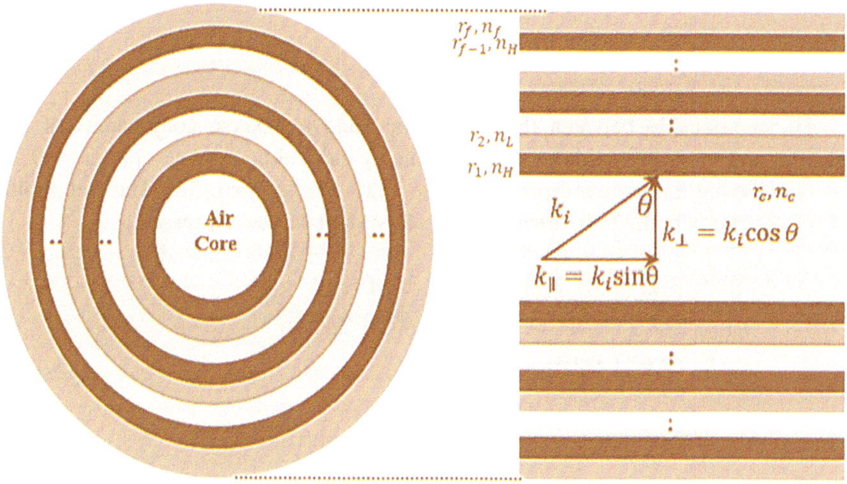

Fig. 3.2 Schematic cross-sectional view and refractive index profile of proposed Bragg Fiber

Let chosen materials be nonmagnetic layers, and their respective refractive indices and thickness are n_H, d_H, n_L, and d_L. The optical path length of the electromagnetic waves in various periodic layers is intended to be one-fourth of a wavelength and can be determined using the following: $n_H d_H - n_L d_L = \lambda c/4$, where λ_c is the design wavelength of the Bragg fiber. The chosen index profile for the whole Bragg structure is given as follows:

$$n(r) = \begin{cases} n_c, & 0 \le r < r_c; \\ n_H, & r_c \le r < r_1; \\ n_L, & r_1 \le r < r_2; \\ \vdots & \\ \vdots & \\ n_H, & r_n \le r < r_{n+1}; \\ n_L, & r_{n+1} \le r < r_{n+2}; \\ \vdots & \\ \vdots & \\ \text{etc.} & \end{cases} \tag{3.12}$$

The reflection and transmission characteristics of a Bragg multilayer fiber can be determined using the cylindrical wave transfer matrix [36, 37]. The Maxwell's equations for a layer with permittivity ε and permeability μ are given by equation [38]:

$$\nabla \times E = -j\omega\mu H, \tag{3.13a}$$

$$\nabla \times H = j\omega\varepsilon E. \tag{3.13b}$$

The relationship between the longitudinal and transverse components of the electric and magnetic fields is formed in the cylindrical coordinate system (r, θ, z) by considering the temporal part of all the fields as exp ($j\omega t$) [38]. The Helmholtz wave equations in the cylindrical coordinate system can be expressed as

$$r\frac{\partial}{\partial r}\left(r\frac{\partial H_z}{\partial r}\right) - r^2\frac{1}{\varepsilon}\frac{\partial \varepsilon}{\partial r}\frac{\partial H_z}{\partial r} + \frac{\partial}{\partial \theta}\left(\frac{\partial H_z}{\partial \theta}\right) + \omega^2\mu\varepsilon r^2 H_z = 0, \text{ (For H-polarization)} \tag{3.14a}$$

$$r\frac{\partial}{\partial r}\left(r\frac{\partial E_z}{\partial r}\right) - r^2\frac{1}{\mu}\frac{\partial \mu}{\partial r}\frac{\partial E_z}{\partial r} + \frac{\partial}{\partial \theta}\left(\frac{\partial E_z}{\partial \theta}\right) + \omega^2\mu\varepsilon r^2 E_z = 0, \text{ (For E-polarization)} \tag{3.14b}$$

The aforementioned equations can be resolved through the utilization of the variable separation technique. The solution for cylindrical waves can be obtained by independently deriving the components for the radial and angular sections. Therefore, the Hz field component of an H-polarized wave can be expressed as:

$$H_z = F(r)\Theta(\theta) \text{ with } \Theta(\theta) \approx e^{im\theta}, \tag{3.15}$$

where m is an integer. Consider the case where the permittivity is radially uniform. Then, $\partial\varepsilon/\partial r = 0$, and radial part of the Eq. (3.13) can be written as:

$$r\frac{d}{dr}\left(r\frac{dF}{dr}\right) + (k^2r^2 - m^2)F = 0,, \tag{3.16}$$

The solution to the aforementioned conventional Bessel's equation is as follows:

$$F(r) = AJ_m(kr) + BY_m(kr), \tag{3.17}$$

where J_m is a Bessel function, A and B are any arbitrary constants, Y_m is a Neumann function and $k = (\omega/c)n$ is wave vector in a medium with refractive index n is denoted by k, while c represents the speed of light in a vacuum. In the scenario of oblique incidence occurring at a core-cladding interface, the wave vector in question will be the component of the incident wave vector that is perpendicular to the interface, as depicted in Fig. 3.2. and is given by $k = k_i$ cos (θ). The incidence wave vector, denoted as k_i, is the vector representing the direction and magnitude of the incident wave that strikes the interface between the core and cladding at an oblique angle θ relative to the perpendicular line drawn on the interface. The magnetic field can be expressed by utilizing the Eqs. (3.16) and (3.17) in conjunction:

$$H_z(r,\theta) = [AJ_m(kr) + BY_m(kr)]e^{im\theta}. \tag{3.18a}$$

The expression for the H-polarization wave, corresponding to nonzero electric fields, can be represented as:

$$E_r = \frac{m}{\omega\varepsilon}\frac{F(r)}{r}e^{im\theta}. \tag{3.18b}$$

$$E_\theta = \frac{1}{-j\omega\varepsilon}\frac{dF}{dr}e^{im\theta} = U(r)e^{im\theta}. \tag{3.18c}$$

$$\frac{dF}{dr} = -j\omega\varepsilon U(r). \tag{3.18d}$$

Using Eq. (3.18d) $U(r)$ can be calculated as:

$$U(r) = jp\left(AJ'_m(kr) + BY'_m(kr)\right), \tag{3.19}$$

where $p = \sqrt{\mu/\varepsilon}$ is the intrinsic impedance of the medium ($p = 377\ \Omega$ for free space).

3.3.1 1-D Bloch Theorem for Bragg Fibers (BFs)

Within a periodic arrangement, the wave undergoes Bragg diffraction, a fundamental physical phenomenon. The Bloch theorem is commonly utilized to examine the wave function in a stratified material. The theorem is specifically intended for application in the Cartesian coordinate system. The Bloch wavenumber is determined by imposing a long enough periodicity in accordance with the periodic boundary condition. A Bragg fiber, specifically engineered for application in fiber optics, comprises of a vacant core and a cladding that periodically alternates between high and low refractive indices. The primary objective of this chapter is to analyze the Bloch theorem specifically in the context of cylindrical coordinates. To start, we shall present a description of the Bloch theorem in cylindrical coordinates. Assuming the presence of an electromagnetic field propagating in the z direction within a material, with an angular frequency ω, β denotes the propagation constant of the characteristic parameters of optical fibers.

With a spatiotemporal factor $U_{tz} = \exp[i(\omega t - \beta z)]$, the field component parallel to the z direction is separated into

$$\psi_z = U_{tz} F(r) \exp(i\upsilon\theta) \tag{3.20}$$

where υ is the azimuthal mode number. The function $F(r)$ fulfills the well-known Bessel differential equation, which is equivalent to the Maxwell equation in cylindrical polar coordinates:

$$\frac{d^2F(r)}{dr^2} + \frac{1}{r}\frac{dF(r)}{dr} + \left[\kappa^2(r) - \frac{\upsilon^2}{r^2}\right]F(r) = 0 \tag{3.21}$$

where $\kappa(r)$ is the lateral propagation constant. Introducing a function $\Phi(r) \equiv \sqrt{r}\,F(r)$, we determine that the equation mentioned above can be formally simplified to the equation of a harmonic oscillator,

$$\frac{d^2\Phi(r)}{dr^2} + \kappa^2(r)\Phi(r) = 0 \tag{3.22}$$

The material is presumed to possess a periodic structure in the r direction, characterized by $\kappa\ (r + \Lambda) = \kappa(r)$, where Λ represents the period. Within this particular context, both $\Phi(r)$ and $\Phi(r + \Lambda)$ adhere to the identical Eq. (3.22), which is

$$\Phi(r + \Lambda) = C(\Lambda)\Phi(r) \tag{3.23}$$

where $C\ (\Lambda)$ is a coefficient dependent on the period Λ. Repeating the use of Eq. (3.23) results in a relationship $C(N\Lambda) = [C(\Lambda)]^N$ where N is any arbitrary integer. This interaction enables us to articulate $C(\Lambda) = \exp(-iK\Lambda)$ with a complex number K. If the system exhibits the periodic boundary condition with a sufficiently long periodicity, such as in the scenario of a single-dimensional system, then $C(N\Lambda) = 1$.

In the present cylindrical system that is bounded at the origin, we obtain

$$\Phi(r + \Lambda) = \exp(-\,iK\Lambda)\Phi(r) \tag{3.24}$$

where the Bloch number K is a complex number. The relationship implies unequivocally that $\Phi(r)$ should be represented by

$$\Phi(r) = \exp(-\,iKr)u(r) \tag{3.25}$$

where $u(r)$ is a periodic function, namely $u(r + \Lambda) = u(r)$. The Bloch theorem is mathematically formulated to describe the behavior of a cylindrical electromagnetic field as it propagates through a periodic material,

$$\sqrt{r+\Lambda}\,F(r+\Lambda) = \exp\,(-\,iK\Lambda)\ \sqrt{r}F(r), \tag{3.26}$$

$$F(r) = \exp(-iKr)\frac{u(r)}{\sqrt{r}} \tag{3.27}$$

It should be noted that $F(r)$ exhibits a singularity at $r = 0$. The reason for this is that in the cylindrical form, the periodic structure is bounded at the origin, while the original Bloch theorem assumes an unbounded periodic structure. Nevertheless, the region surrounding the origin lies beyond the cladding area, making it unsuitable for the application of the Bloch theorem. This theorem ensures that a specific core radius is required for the propagation mode, in accordance with the guiding limit.

3.3.2 *Mathematical Hankel Formalism*

The field solution for the cylindrical wave can be obtained by deriving the radial and angular components in the preceding section. For a material that is uniformly distributed radially, this solution can be employed to analyze the properties of wave propagation. However, when dealing with a nonuniform medium, it is more suitable to regard the electromagnetic wave as the combination of two opposing waves that propagate in opposite directions. This involves the superposition of incoming waves (converging) and outgoing waves (diverging). The two cylindrical waves are denoted by two Hankel functions. The incoming cylindrical wave that is monochromatic and H-polarized can be described as follows:

$$H_z^- = BH_m^{(1)}(kr)\exp(jm\theta) \tag{3.28a}$$

$$E_\theta^- = jpBH_m^{(1)\prime}(kr)\exp(jm\theta), \tag{3.28b}$$

where $H_m^{(1)}$ is the Hankel function of the first kind. The Hankel function of the second kind $H_m^{(2)}$ represents the outward wave,

$$H_z^+ = AH_m^{(2)}(kr)\exp(jm\theta), \tag{3.28c}$$

$$E_\theta^+ = jpAH_m^{(2)\prime}(kr)\exp(jm\theta), \tag{3.28d}$$

where m is the azimuthal variation of a field, and A and B are any arbitrary constant, $p = \sqrt{\mu/\varepsilon}$ which is known as the intrinsic impedance of the medium. The field component H_z E_θ can be written as,

$$H_z = H_z^+ + H_z^-. \tag{3.29a}$$

$$E_\theta = E_\theta^+ + E_\theta^- = jp\left[C_m^{(2)}H_z^+ + C_m^{(1)}H_z^-\right], \tag{3.29b}$$

where $C_{ml}^{(1,2)} = H_m^{(1,2)\prime}(kr)/H_m^{(1,2)}(kr)$, with l is the respective layer.

3.3.3 *Optical Properties of Bragg Fibers (BFs)*

Propagation Wave Essential Functions

The cylindrical wave's field solution can be divided into radial and angular components, as seen in the preceding section. The radial component is characterized by the Bessel function J_m and the Neumann function Y_m. However, when dealing with the issue of wave propagation, it is advantageous to represent the field solution as the combination of two opposing waves that propagate in opposite directions. This means that the solution is formed by adding together waves that converge towards a point and waves that diverge away from it. The two waves are commonly depicted using two Hankel functions. The H-polarization is characterized by cylindrical waves with magnetic and electric fields that have a specific form,

$$H_z^+ = AH_m^{(2)}(kr)\exp(jm\theta), \tag{3.30a}$$

$$E_\theta^+ = jpAH_m^{(2)\prime}(kr)\exp(jm\theta), \tag{3.30b}$$

where $H_m^{(2)}$ is the Hankel function of the second kind. On the other hand, the ingoing wave is Hankel function of the first kind, namely

$$H_z^- = BH_m^{(1)}(kr)\exp(jm\theta), \tag{3.30c}$$

$$E_\theta^- = jpBH_m^{(1)\prime}(kr)\exp(jm\theta), \tag{3.30d}$$

For a field with azimuthal variation specified by "m," the total field of both H_z and E_θ can be written as:

$$H_z = H_z^+ + H_z^- \tag{3.31}$$

and

$$E_\theta = E_\theta^+ + E_\theta^- = jpC_m^{(2)}H_z^+ + jpC_m^{(1)}H_z^- \tag{3.32}$$

where

$$C_m^{(1,2)} = \frac{H_m^{(1,2)'}(kr)}{H_m^{(1,2)}(kr)} \tag{3.33}$$

To denote the layer in which the field is present, we insert the layer label as a secondary subscript on the coefficient $C_m^{(1,2)}$, and as a subscript of the wave vector k, namely

$$C_m^{(1,2)} = \frac{H_m^{(1,2)'}(k_l r)}{H_m^{(1,2)}(k_l r)}, \tag{3.34}$$

for a layer, say l.

Next, we would like to relate the magnetic fields at the two boundaries for a single layer. The relationship can be described by a matrix **P** which plays a similar role as the propagation matrix in a standard TMM in usual planar geometry [26]. This matrix can be constructed as follows. From Eq. (3.30), the outgoing magnetic field is rewritten by

$$H_z^+(r) = \frac{H_m^{(2)}(kr)}{H_m^{(2)}(kr_0)} A H_m^{(2)}(kr_0) \exp(jm\theta) = \frac{H_m^{(1)}(kr)}{H_m^{(1)}(kr_0)} H_z^+(r_0), \tag{3.35a}$$

and the ingoing magnetic field is expressed as

$$H_z^-(r) = \frac{H_m^{(1)}(kr)}{H_m^{(1)}(kr_0)} A H_m^{(1)}(kr_0) \exp(jm\theta) = \frac{H_m^{(1)}(kr)}{H_m^{(1)}(kr_0)} H_z^-(r_0) \tag{3.35b}$$

Equations (3.35a) and (3.35b) can be combined as a matrix equation, namely

$$\begin{pmatrix} H_z^+(r) \\ H_z^-(r) \end{pmatrix} = P \begin{pmatrix} H_z^+(r_0) \\ H_z^-(r_0) \end{pmatrix}, \tag{3.36}$$

where the propagation matrix is diagonal and can be expressed as

$$P = \begin{pmatrix} \dfrac{H_m^{(2)}(kr)}{H_m^{(2)}(kr_0)} & 0 \\ 0 & \dfrac{H_m^{(1)}(kr)}{H_m^{(1)}(kr_0)} \end{pmatrix} \tag{3.37}$$

This matrix transforms the magnetic field from the inner boundary, where "r = r$_0$," to the point "r = r" within the layer. It is important to note that the matrix "P" for "E-polarization" is identical to the matrix "P" stated.

In the conventional two-dimensional transfer matrix method (TMM), there exists a significant matrix known as the dynamical matrix "D," which establishes the connection between the boundary conditions of two different media [26]. Furthermore, a comparable matrix is also present in the cylindrical multilayer system. Let's examine the interface between two layers by focusing on the consistency of the tangential components of the electric and magnetic fields (H_z, E_θ). The interface condition can be formulated in terms of $\left(H_z^+, H_z^-\right)$. Now, we present matrix D, which transforms the basis $\left(H_z^+, H_z^-\right)$ to (H_z, E_θ) namely:

$$\begin{pmatrix} H_z \\ E_\theta \end{pmatrix} = D \begin{pmatrix} H_z^+ \\ H_z^- \end{pmatrix}, \tag{3.38}$$

Where D is written by

$$D = \begin{pmatrix} 1 & 1 \\ jpC_m^{(2)} & jpC_m^{(1)} \end{pmatrix} \tag{3.39}$$

For the nonmagnetic medium, $\mu_r = 1$, the determinant of D is found to be;

$$detD = -\frac{1}{\varepsilon_r}\sqrt{\frac{\mu_0}{\varepsilon_0}}\frac{4}{\pi Kr}\frac{1}{H_m^{(2)}(kr)H_m^{(1)}(kr)}, \tag{3.40}$$

where the wave number of free space, denoted as K, is defined as the ratio of the angular frequency (ω) to the speed of light (c) and, in arriving at Eq. (3.40), we have used the following Wronskian;

$$W\left[H_\nu^{(1)}(z), H_\nu^{(2)}(z)\right] = H_\nu^{(1)}(z)H_\nu^{(2)\prime}(z) - H_\nu^{(1)\prime}(z)H_\nu^{(2)}(z) = -\frac{4j}{\pi z}. \tag{3.41}$$

Also, the inverse of D can be calculated to be:

$$D^{-1} = \varepsilon_r\sqrt{\frac{\varepsilon_0}{\mu_0}}\frac{\pi}{4}KrH_m^{(2)}(kr)H_m^{(1)}(kr)\begin{pmatrix} -jpC_m^{(1)} & 1 \\ jpC_m^{(2)} & -1 \end{pmatrix}. \tag{3.42}$$

Next, given the requirement for the tangential field components to be continuous at the interface between two layers labeled 1 and 2, we have

$$D_1\begin{pmatrix} H_{z1}^+ \\ H_{z1}^- \end{pmatrix} = D_2\begin{pmatrix} H_{z2}^+ \\ H_{z2}^- \end{pmatrix} \tag{3.43}$$

Equations (3.43) can be rewritten by

$$\begin{pmatrix} H_{z2}^+ \\ H_{z2}^- \end{pmatrix} = D_2^{-1}D_1\begin{pmatrix} H_{z1}^+ \\ H_{z1}^- \end{pmatrix} = D_{21}\begin{pmatrix} H_{z1}^+ \\ H_{z1}^- \end{pmatrix} = \begin{pmatrix} d_{11} & d_{12} \\ d_{21} & d_{22} \end{pmatrix}\begin{pmatrix} H_{z1}^+ \\ H_{z1}^- \end{pmatrix} \tag{3.44}$$

The matrix $D_{21} = D_2^{-1}D_1$ is considered a transmission matrix that connects the amplitudes of the waves on both sides of the interface. By utilizing Eqs. (3.39) and (3.42), one may get the matrix element of D_21, yielding the following outcomes:

$$d_{11} = -j\varepsilon_r\sqrt{\frac{\varepsilon_0}{\mu_0}}\frac{\pi}{4}K\rho H_m^{(2)}(k_2\rho)H_m^{(1)}(k_2\rho)\left[p_2C_m^{(1)} - p_1C_m^{(2)}\right], \tag{3.45a}$$

$$d_{21} = -j\varepsilon_r\sqrt{\frac{\varepsilon_0}{\mu_0}}\frac{\pi}{4}K\rho H_m^{(2)}(k_2\rho)H_m^{(1)}(k_2\rho)\left[p_1C_m^{(2)} - p_2C_m^{(2)}\right], \tag{3.45b}$$

$$d_{11} = -j\varepsilon_r\sqrt{\frac{\varepsilon_0}{\mu_0}}\frac{\pi}{4}K\rho H_m^{(2)}(k_2\rho)H_m^{(1)}(k_2\rho)\left[p_2C_m^{(1)} - p_1C_m^{(2)}\right], \tag{3.45c}$$

$$d_{11} = -j\varepsilon_r\sqrt{\frac{\varepsilon_0}{\mu_0}}\frac{\pi}{4}K\rho H_m^{(2)}(k_2\rho)H_m^{(1)}(k_2\rho)\left[p_2C_m^{(1)} - p_1C_m^{(2)}\right]. \tag{3.45d}$$

Optical Properties for Single Cylindrical Interface

The equation describing the reflection and transmission of waves between two different mediums is similar to the well-known Fresnel's equations used in planar geometry. Let's examine a wave that is spreading outwards and approaching the interface between layers 1 and 2. The reflection coefficient, denoted as r_d, and the transmission coefficient must satisfy the following relation:

$$\begin{pmatrix} t_d \\ 0 \end{pmatrix} = D_{21}\begin{pmatrix} 1 \\ r_d \end{pmatrix} \tag{3.46}$$

Equation (3.46) reveals that r_d and t_d are expressed as follows:

$$r_d = -\frac{d_{21}}{d_{22}} = \frac{p_2C_{m2}^{(2)} - p_1C_{m1}^{(2)}}{p_1C_{m1}^{(1)} - p_2C_{m2}^{(2)}}, \tag{3.47}$$

and

$$t_d = d_{11} + d_{12}r_d = d_{11} - d_{12}\frac{d_{21}}{d_{22}} = \frac{detD_{21}}{d_{22}} \tag{3.48}$$

The determinant of D_{21}can be computed using Eqs. (3.39) and (3.42), namely

$$detD_{21} = detD_2^{-1}detD_1$$

$$= \varepsilon_r\sqrt{\frac{\varepsilon_0}{\mu_0}}\left[\frac{\pi}{4}K\rho H_m^{(2)}(k_2\rho)H_m^{(1)}(k_2\rho)\right]^2\begin{vmatrix} -jp_2C_{m2}^{(1)} & 1 \\ -jp_2C_{m2}^{(2)} & 1 \end{vmatrix}\cdot\frac{-4}{\pi K\rho H_m^{(2)}(k_1\rho)H_m^{(1)}(k_1\rho)} \tag{3.49}$$

Therefore, the transmission coefficient in Eq. (3.48) can be cast as

$$t_d = \frac{-j4\varepsilon_r^{-1}\sqrt{\frac{\mu_0}{\varepsilon_0}}}{\pi K\rho H_m^{(2)}(k_1\rho)H_m^{(1)}(k_1\rho)\left[p_2C_{m2}^{(2)} - p_1C_{m1}^{(1)}\right]} \tag{3.50}$$

Equations (3.47) and (3.50) provide the reflection coefficient and transmission coefficient, respectively, for the diverging wave. Similarly, the reflection coefficient and transmission coefficient for the converging wave can be calculated based on the observed values.

$$r_c = \frac{p_2C_{m2}^{(1)} - p_1C_{m2}^{(1)}}{p_1C_{m1}^{(1)} - p_2C_{m2}^{(2)}}, \tag{3.51}$$

$$t_c = \frac{-j4\varepsilon_r^{-1}\sqrt{\frac{\mu_0}{\varepsilon_0}}}{\pi K\rho H_m^{(2)}(k_2\rho)H_m^{(1)}(k_2\rho)\left[p_2C_{m2}^{(2)} - p_1C_{m1}^{(1)}\right]} \tag{3.52}$$

The Eqs. (3.47), (3.50), (3.51), and (3.52) define the reflection coefficient and transmission coefficient. These equations provide the Fresnel-like equations for the cylindrical interface between media 1 and 2. Furthermore, these four equations exhibit identical structures to those in the E-polarization, but with a distinct parameter "p" specification.

Reflection and Transmission in Cylindrical Slab

Now, let's direct our attention to the phenomenon of wave reflection and transmission occurring in a solitary cylindrical slab. This matter bears resemblance to the situation involving the Airy formula for a solitary slab in the conventional planar configuration. The picture below (Fig. 3.3) illustrates a single layer with an index of refraction n_1 and a thickness of $r_1 - r_0$. For illustrative purposes, we have plotted a straight line to represent the cylindrical contact. The reflection coefficient "r" and transmission coefficient "t" can be obtained by following these steps. At $=r_1$, i.e., the second interface, let us consider the transmission and reflection coefficients as t_{1d}

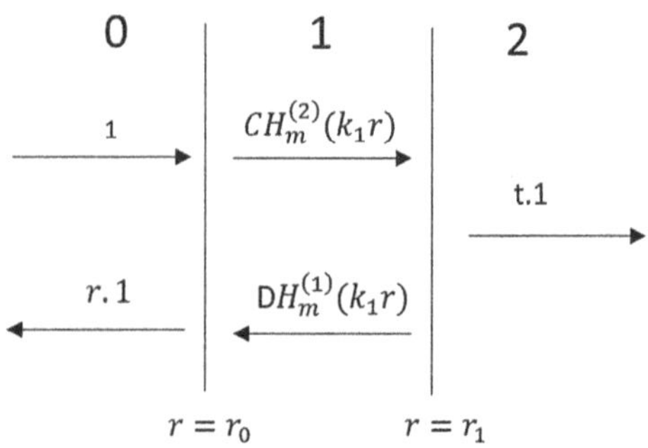

Fig. 3.3 A single circular layer of index n_1 and of thickness "$r_1 - r_0$" bounded by media "0" and "2." The curved interfaces are deliberately shown as straight lines for the sake of clarity in the illustration

and r_{1d} respectively. Additionally, we shall consider the transmission and reflection coefficient of the first interface $r = r_0$ as t_{0d}, t_{0c}, r_{0d} and r_{0c}. The subscript "d" denotes the outgoing wave, whereas the subscript "c" represents the approaching wave.

The formula for "r" and "t" is provided explicitly as:

$$r = \frac{r_{0d} + (t_{0c}t_{0d} - r_{0c}r_{1d})r_{1d}\Theta}{1 - r_{0c}r_{1d}\Theta}, \tag{3.53}$$

$$t = \frac{t_{0d}t_{1d}}{1 - r_{0c}r_{1d}\Theta} \frac{H_m^{(2)}(k_1r_1)}{H_m^{(2)}(k_1r_0)}, \tag{3.54}$$

where

$$\Theta = \frac{H_m^{(1)}(k_1r_1)H_m^{(2)}(k_1r_1)}{H_m^{(2)}(k_1r_1)H_m^{(1)}(k_1r_1)}, \tag{3.55}$$

and k_1 is the wave vector in layer 1.

The derivation of Eqs. (3.53, 3.54, and 3.55) can be described as follows. At the first boundary $r = r_0$, we have

$$CH_m^{(2)}(k_1r_0) = 1.t_{0d} + r_{0c}.DH_m^{(1)}(k_1r_0), \tag{3.56}$$

$$1.r = 1.r_{0d} + t_{0c} + r_{0c}.DH_m^{(1)}(k_1r_0), \tag{3.57}$$

In a similar fashion, we can write at the second boundary $r = r_1$,

$$t.1 = t_{1d}.CH_m^{(1)}(k_1r_1), \tag{3.58}$$

$$DH_m^{(1)}(k_1r_1) = r_{1d}.CH_m^{(2)}(k_1r_0), \tag{3.59}$$

Eliminating D from Eqs. (3.56) and (3.57) lead to

$$C\left[H_m^{(2)}(k_1r_0) - r_{1d}\frac{H_m^{(2)}(k_1r_1)}{H_m^{(1)}(k_1r_1)}t_{0c}H_m^{(1)}(k_1r_0)\right] = t_{0d} \tag{3.60}$$

Again, eliminating D from Eqs. (3.57) and (3.59) give

$$r - r_{0d} = Cr_{1d}\frac{H_m^{(2)}(k_1r_1)}{H_m^{(1)}(k_1r_1)}t_{0c}H_m^{(1)}(k_1r_0) \tag{3.61}$$

By referring to Eqs. (3.60) and (3.61), we can straightforwardly obtain the reflection coefficient as stated in Eq. (3.47). By utilizing Eqs. (3.56) and (3.59), we may obtain the transmission coefficient as stated in Eq. (3.50). Furthermore, Eqs. (3.53, 3.54 and 3.55) exhibit identical characteristics to those seen in the E-polarization.

Reflection and Transmission in a Cylindrical Multilayer Structure

In the preceding part, we have examined the mathematical formula that describes the reflection and transmission coefficients for the propagation of waves in a cylindrical multilayer structure. Expanding on these equations, we will analyze the characteristics of waves in a periodic cylindrical multilayer structure, commonly referred to as a cylindrical Bragg reflector. With respect to Fig. 3.2, there is a wave that is moving away from the source and is expected to reach the interface located at $r = r_0$, between 0 and 1. From there, it continues to propagate into the final medium f, which is anticipated to extend from $r = r_f to\ r = \infty$. The magnitudes of the magnetic field and electric fields at r_0 and r_f can be expressed in relation to the amplitude reflection and transmission coefficients $'r_d'$ and $'t_d'$, along with the transfer matrix T. This yields the following outcomes:

$$T_{11} = \frac{\pi}{2} kr_0 \left[Y'_m(kr_0)J_m(kr) - J'_m(kr_0)Y_m(kr)\right], \tag{3.62a}$$

$$T_{12} = \frac{-j\pi}{2}\frac{k}{p} r_0 [J_m(kr_0)Y_m(kr) - Y_m(kr_0)J_m(kr)], \tag{3.62b}$$

$$T_{21} = j\frac{\pi}{2} kr_0 p \left[Y'_m(kr_0)J'_m(kr) - J'_m(kr_0)Y'_m(kr)\right], \tag{3.62c}$$

$$T_{22} = \frac{\pi}{2} kr_0 \left[J_m(kr_0)Y'_m(kr) - Y_m(kr_0)J'_m(kr)\right]. \tag{3.62d}$$

$$\begin{pmatrix} 1 + r_d \\ jp_o C_{m0}^{(2)} + jp_0 C_{m0}^{(1)} r_d \end{pmatrix} = T^{-1} \begin{pmatrix} t_d \\ jp_f C_{mf}^{(2)} t_d \end{pmatrix} \tag{3.63}$$

The inverse matrix appearing on the right-hand side of Eq. (3.62) is represented as

$$T^{-1} = \begin{pmatrix} T_{11} & T_{12} \\ T_{21} & T_{22} \end{pmatrix}^{-1} = \frac{1}{detT} \begin{pmatrix} T_{22} & -T_{12} \\ -T_{21} & T_{11} \end{pmatrix} \equiv \begin{pmatrix} T'_{11} & T'_{12} \\ T'_{21} & T'_{22} \end{pmatrix} \tag{3.64}$$

The expressions for the reflection coefficient r_d and transmission coefficient t_d can be calculated using Eqs. (3.63) and (3.64).

$$r_d = \frac{\left(T'_{21} - jp_0 C_{m0}^{(2)} T'_{11}\right) + jp_f C_{mf}^{(2)} \left(T'_{22} - jp_0 C_{m0}^{(2)} T'_{12}\right)}{\left(jp_0 C_{m0}^{(1)} T'_{11} - T'_{21}\right) + jp_f C_{mf}^{(2)} \left(jp_0 C_{m0}^{(1)} T'_{12} - T'_{22}\right)}, \tag{3.65}$$

$$t_d = \frac{-4\varepsilon_r^{-1}\sqrt{\frac{\mu_0}{\varepsilon_0}}}{\pi K r_0 H_m^{(2)}(k_0 r_0) H_m^{(1)}(k_0 r_0) \left[\left(jp_0 C_{m0}^{(1)} T'_{11} - T'_{21}\right) + jp_f C_{mf}^{(2)} \left(jp_0 C_{mf}^{(1)} T'_{12} - T'_{22}\right)\right]}. \tag{3.66}$$

Equations (3.65) and (3.66) can be used to analyze the photonic band gap structure in a periodic cylindrical multilayer structure. It is worth mentioning that Eqs. (3.65) and (3.66) can be reduced to those for the E-polarization with a replacement of $j \rightarrow -j$.

Galerkin, asymptotic, and TMM for Bragg fibers are reviewed and compared with all other calculation techniques. The most accurate method is determined to be TMM [39, 40]. The transfer matrix technique is, therefore, the most appropriate of all of these methods for evaluating periodic structures. As a result, it has been used throughout the whole current investigation. Further, Sect. 3.5 describes this approach for cylindrical Bragg structures in depth.

3.4 BFs Theoretical Model and Computational Approach

In this Fig. 3.4, n_c, r_c shows the refractive index and radius of core while n_H, n_L and d_H, d_L show respectively, the refractive indices and the respective thicknesses of high and low index layers, which form bilayer structure. $\Lambda = d_H + d_L$, is the bilayer period thickness. Assume that the field components have a spatial-temporal factor of $U_{tz} = \exp\left[i(\omega t - \beta z)\right]$, β being the propagation constant. A multilayer Bragg fiber can be created by repeating these bilayers. The electromagnetic wave in a Bragg fiber waveguide is expected to have a cylindrical shape, which is a result of the cylindrical symmetry of the structure. The cylindrical wave is presumed to be radiating away

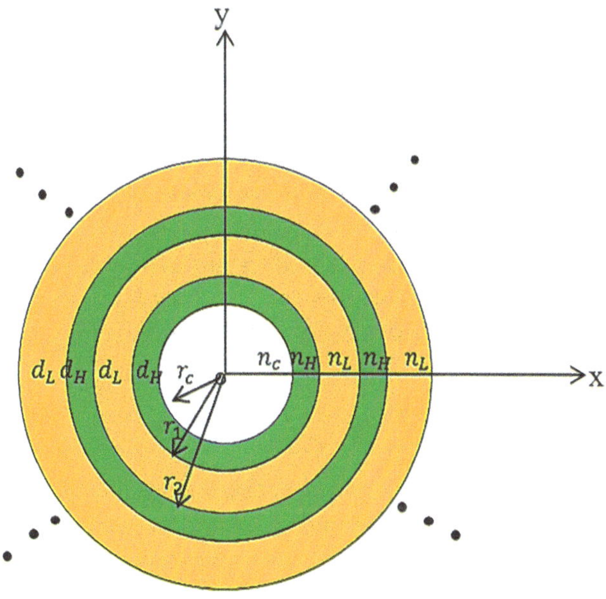

Fig. 3.4 Illustrates a cross-sectional view of the Bragg fiber (BFs) along its length

from the axis of symmetry A, and then impinges against the initial cylindrical contact B. Now, we can utilize the transfer matrix approach for cylindrical waves [36] to determine the reflectance of the entire structure.

It is assumed that the temporal part of all the fields is exp($i\omega t$). Hence, the source free Maxwell's equations [37] are written as;

$$\nabla \times E = -j\omega H, \tag{3.67a}$$

$$\nabla \times H = j\omega E. \tag{3.67b}$$

For TE mode or H polarization in cylindrical coordinates (r, θ, z) [37], the nonzero fields E_z, H_θ, H_r satisfy the following three equations in each layer,

$$\frac{1}{r}\frac{\partial E_z}{\partial \theta} = -j\omega\mu H_r, \tag{3.68a}$$

$$\frac{\partial E_z}{\partial r} = j\omega\mu H_\theta, \tag{3.68b}$$

$$\frac{\partial (rH_\theta)}{\partial r} - \frac{\partial H_r}{\partial \theta} = -j\omega\varepsilon r E_z. \tag{3.68c}$$

Equation (3.68c) can be utilized to express the governing differential equation for the tangential component of the electric field E_z,

$$r\frac{\partial}{\partial r}\left(r\frac{\partial E_z}{\partial r}\right) - r^2\frac{1}{\mu}\frac{\partial \mu}{\partial r}\frac{\partial E_z}{\partial r} + \frac{\partial}{\partial \theta}\left(\frac{\partial E_z}{\partial \theta}\right) + \omega^2\varepsilon\mu r^2 E_z = 0. \tag{3.69}$$

Now, let us assume the field component $E_z = U(r)\Theta(\theta) = U(r)e^{im\theta}$, and the solution U is given as;

$$U(r) = C_1 J_m(kr) + C_2 Y_m(kr), \tag{3.70}$$

where C_1 and C_2 are constants, J_m is a Bessel function, Y_m is the Neumann function and k is the wave number in the material medium. In this investigation, the azimuthal mode number, denoted by the subscript "m," has been set to zero. Similarly, the magnetic field's tangential component can be represented as, $H_\theta = V(r)e^{im\theta}$, where $V(r)$is given by,

$$V(r) = -jp\left(C_1 J'_{\mathrm{m}}(kr) + C_2 Y'_{\mathrm{m}}(kr)\right), \tag{3.71}$$

where $p = \sqrt{\varepsilon/\mu}$ is the intrinsic admittance of the material medium.

Equations (3.70) and (3.71) are used together to create a single layer matrix that connects the electric and magnetic fields at the two interfaces. For instance, the matrix $\widehat{T}_1$ for the first layer at interface ($r = r_0$ and r_1) is written as [36, 37].

$$\begin{bmatrix} U(r_1) \\ V(r_1) \end{bmatrix} = \widehat{T}_1 \begin{bmatrix} U(r_0) \\ V(r_0) \end{bmatrix}. \tag{3.72}$$

The component of the single layer matrix $\widehat{T}_1 = \begin{bmatrix} t_{11} & t_{12} \\ t_{21} & t_{22} \end{bmatrix}$ can be written as

$$t_{11} = \frac{\pi}{2} k_1 r_0 \left[Y'_{\mathrm{m}}(k_1 r_0) J_{\mathrm{m}}(k_1 r_1) - J'_{\mathrm{m}}(k_1 r_0) Y_{\mathrm{m}}(k_1 r_1) \right], \tag{3.73a}$$

$$t_{12} = j \frac{\pi}{2} \frac{k_1}{p_1} r_0 [J_{\mathrm{m}}(k_1 r_0) Y_{\mathrm{m}}(k_1 r_1) - Y_{\mathrm{m}}(k_1 r_0) J_{\mathrm{m}}(k_1 r_1)], \tag{3.73b}$$

$$t_{21} = -j \frac{\pi}{2} k_1 r_0 p_1 \left[Y'_{\mathrm{m}}(k_1 r_0) J'_{\mathrm{m}}(k_1 r_1) - J'_{\mathrm{m}}(k_1 r_0) Y'_{\mathrm{m}}(k_1 r_1) \right], \tag{3.73c}$$

$$t_{22} = \frac{\pi}{2} k_1 r_0 \left[J_{\mathrm{m}}(k_1 r_0) Y'_{\mathrm{m}}(k_1 r_1) - Y_{\mathrm{m}}(k_1 r_0) J'_{\mathrm{m}}(k_1 r_1) \right], \tag{3.73d}$$

where $p_1 = \sqrt{\varepsilon_1/\mu_1}$.

Equation (3.73) demonstrates that the matrix elements are inherently contingent on the radii of the two interfaces. Therefore, one may easily create the matrix representing all the layers of a multilayered Bragg fiber waveguide. In order to generate the total transfer matrix $\widehat{T}$that relates the initial and final interfaces of the Bragg fiber waveguide construction, a total of 2 N matrices are needed. Hence, the aforementioned interfaces are related as,

$$\begin{bmatrix} U(r_{\mathrm{f}}) \\ V(r_{\mathrm{f}}) \end{bmatrix} = \widehat{T} \begin{bmatrix} U(r_0) \\ V(r_0) \end{bmatrix}, \tag{3.74}$$

where

$$\widehat{T} = \begin{bmatrix} T_{11} & T_{12} \\ T_{21} & T_{22} \end{bmatrix} = \widehat{T}_{2\mathrm{N}} \widehat{T}_{2\mathrm{N}-1} \ldots \ldots \widehat{T}_{\mathrm{N}+1} \widehat{T}_{\mathrm{N}} \ldots \ldots \widehat{T}_2 \widehat{T}_1. \tag{3.75}$$

The reflection and transmission coefficients can be obtained using the calculated transfer matrix by using the following relationships [36],

$$r_{\mathrm{d}} = \frac{\left(T'_{21} + jp_0 C^{(2)}_{\mathrm{m0}} T'_{11}\right) - jp_f C^{(2)}_{\mathrm{mf}} \left(T'_{22} + jp_0 C^{(2)}_{\mathrm{m0}} T'_{12}\right)}{\left(-jp_0 C^{(1)}_{\mathrm{m0}} T'_{11} - T'_{21}\right) - jp_{\mathrm{f}} C^{(2)}_{\mathrm{mf}} \left(-jp_0 C^{(1)}_{\mathrm{m0}} T'_{12} - T'_{22}\right)}, \tag{3.76}$$

$$t_{\mathrm{d}} = \frac{4\sqrt{\varepsilon_0/\mu_0}}{\pi K r_0 H^{(2)}_{\mathrm{m}}(k_0 r_0) H^{(1)}_{\mathrm{m}}(k_0 r_0) \left[\left(-jp_0 C^{(1)}_{\mathrm{m0}} T'_{11} - T'_{21}\right) - jp_{\mathrm{f}} C^{(2)}_{\mathrm{mf}} \left(-jp_0 C^{(1)}_{\mathrm{m0}} T'_{12} - T'_{22}\right) \right]}, \tag{3.78}$$

where $p_0 = \sqrt{\varepsilon_0/\mu_0}$ and $p_f = \sqrt{\varepsilon_f/\mu_f}$ are the admittance of core and final media, $T'_{11}, T'_{12}, T'_{21}$ and T'_{22} are the matrix element of the inverse matrix of $\hat{T}$, $K = \omega\sqrt{\varepsilon_0\mu_0}$ is the free-space wave number and

$$C_{\mathrm{ml}}^{(1,2)} = \frac{H_{\mathrm{m}}^{(1,2)\prime}(k_l r_l)}{H_{\mathrm{m}}^{(1,2)}(k_l r_l)}, l = 0, f. \quad (3.79)$$

where $H_m^{(1)}$ and $H_m^{(2)}$ are the Hankel function of the first and second kind. Using Eqs. (3.76) and (3.77), the percentage reflectance $\%R$ and Transmittance $\%T$ are:

$$\%R = 100 \times |r_d|^2, \%T = 100 \times \frac{n_{\mathrm{out}}}{n_c}|t_d|^2. \quad (3.80)$$

where n_c and n_{out} are the refractive indices of the core and outer media.

The analysis of the proposed waveguide structures [41–46] can be conducted by utilizing several sensing metrics, including sensitivity, detection accuracy, and quality parameter, which are crucial factors. The sensitivity of a sensor is determined by the ratio of the measured output of the sensor to the change in the physical quantity being measured. The mathematical definition of sensor sensitivity is the ratio of the spectral shift in resonance wavelength to the small change in refractive index of the aqueous medium or analyte contained in the core.

$$S = \Delta\lambda_{\mathrm{res}}/\Delta n_c. \quad (3.81)$$

Detection accuracy refers to the degree of precision in determining the true value of the measurand. The sensor's detection accuracy is determined by the ratio of the shift in resonance wavelength to the full width at half maximum (FWHM) of the photonic band gap. The quality parameter provides an assessment of the sensor's performance by considering its sensitivity and accuracy in detecting signals.

3.5 Hankel Formulation and Characteristic Equation for BFs Device Structure

Figure 3.5 displays the schematic refractive index profile of the waveguide. This suggested waveguide exhibited propagation of both TE and TM modes. However, our current study specifically examines the TE mode, which exhibits nonzero field components of H_r, E_θ *and* H_z. Additional elements can also be obtained from these components. The electromagnetic fields are composed of fundamental components that are chosen to be continuous at the interface of each layer. Accordingly, H_z *and* E_θ are [47],

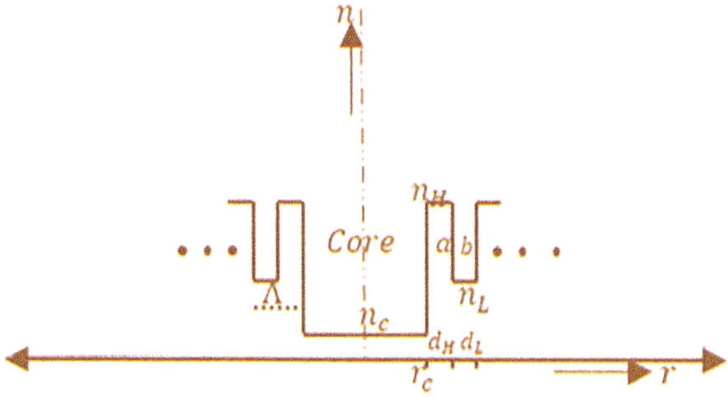

Fig. 3.5 Schematic diagram of refractive index profile with radial position of core and thickness

$$\begin{pmatrix} H_z \\ iE_\theta \end{pmatrix} = U_{tz} D_i^{TE}(r) \begin{pmatrix} A_i \\ B_i \end{pmatrix} \tag{3.82}$$

where, $D_i^{TE}(r) \equiv \begin{pmatrix} d_{11}^{TE} & d_{12}^{TE} \\ d_{21}^{TE} & d_{22}^{TE} \end{pmatrix}$ is referred to as the representation matrix. Within the core region, every element of the representation matrix is defined as follows,

$$d_{11}^{TE} = H_\nu^{(2)}(\kappa_c r), \tag{3.83a}$$

$$d_{12}^{TE} = H_\nu^{(1)}(\kappa_c r), \tag{3.83b}$$

$$d_{21}^{TE} = -\left(\frac{\omega \mu_0}{\kappa_c}\right) H_\nu^{(2)'}(\kappa_c r), \tag{3.83c}$$

$$d_{22}^{TE} = \left(\frac{n_c^2 \omega \varepsilon_0}{\kappa_c}\right) H_\nu^{(1)'}(\kappa_c r). \tag{3.83d}$$

$$\kappa_c \equiv \sqrt{(n_c k_0)^2 - \beta^2}\ ; 0 \le r \le r_c. \tag{3.84}$$

The Hankel functions, $H_\nu^{(1)} = J_\nu + iN_\nu$ and $H_\nu^{(2)} = J_\nu - iN_\nu$ represent the inward and outward waves, respectively. Prime calculates the derivative of the function with respect to its argument. Additionally, k_0, which represents the wavenumber in vacuum, is defined as the ratio of the angular frequency ω to the speed of light c. To express fields in the claddings of the waveguide, which has a rather high core radius, we can utilize an asymptotic extension of the Hankel function at large parameters [47]. The representation matrix in this scenario is displayed using the boundary matrix P_i^{TE} and the displacement matrix Q_i^{TE}.

$$D_i^{TE}(r) \simeq \sqrt{\frac{2}{\pi \rho}} P_i^{TE} Q_i(r); \tag{3.85a}$$

where,

$$P_i^{TE} = \frac{1}{\sqrt{\kappa_i}} \begin{pmatrix} 1 & 1 \\ i\omega\mu_0/\kappa_i & -i\omega\mu_0/\kappa_i \end{pmatrix} \tag{3.85b}$$

$$Q_i^{TE}(\rho) = \begin{pmatrix} \exp(-i\kappa_i r) & 0 \\ 0 & \exp(i\kappa_i r) \end{pmatrix}; \tag{3.85c}$$

The subscript i represents layers a, b, and c. The relationship between the amplitudes of the TE mode for layer "a" can be expressed as,

$$\begin{pmatrix} a_{m+1} \\ b_{m+1} \end{pmatrix} = \begin{pmatrix} X_{TE} & Y_{TE} \\ Y_{TE}^* & X_{TE}^* \end{pmatrix} \begin{pmatrix} a_m \\ b_m \end{pmatrix}; \tag{3.86a}$$

$$X_{TE} = \left[\cos(\kappa_b d_L) - \frac{i}{2}\left(\frac{\kappa_b}{\kappa_a} + \frac{\kappa_a}{\kappa_b}\right) \sin(\kappa_b d_L) \right] \exp(-i\kappa_a d_H) \tag{3.86b}$$

$$Y_{TE} = \frac{i}{2}\left(\frac{\kappa_b}{\kappa_a} - \frac{\kappa_a}{\kappa_b}\right) \sin(\kappa_b d_L) \,] \exp(i\kappa_a d_H) \tag{3.86c}$$

Each coefficient in the above expression satisfies $|X_{TE}|^2 - |Y_{TE}|^2 = 1$ the asymptotic expansion approximation. Now, using the Floquet theorem [47] in the periodic structure, we can find out the eigenvalue equation for the proposed Bragg waveguide. According to Floquet, if the material has a periodicity such as, $n(r + \Lambda) = n(r)$, then the wave propagating in such a medium is expressed as,

$$G_K(r, z) = G_K(r) \exp(-iKr) \exp(-i\beta z) \tag{3.87}$$

In this context, K represents the Bloch wavenumber, which is a constant value applicable to all periodic structures. The eigenvalue equation can be obtained by deriving the boundary condition at the interface between the core and the cladding. It yields an outcome,

$$\frac{J_\nu'(\kappa_c r_c)}{J_\nu(\kappa_c r_c)} + \frac{\kappa_c}{i\kappa_a} \frac{\{[\exp(-iK^{TE}\Lambda) - X_{TE}] - Y_{TE}\}}{\{[\exp(-iK^{TE}\Lambda) - X_{TE}] + Y_{TE}\}} = 0 \tag{3.88}$$

Various characteristics can be evaluated from the above equation.

3.6 Conclusion

The present chapter successfully deals with various analytical techniques, e.g., Asymptotic, Galerkin, and transfer matrix methods to theoretically and also mathematically model the layer-by-layer periodic cylindrical layers of BFs structure. In this book chapter, the utility of transfer matrix techniques over other above existing techniques in layer-by-layer analysis of existing structure. During modeling processes, 1D-cylindrical Bloch theorem for the periodic structure and Hankel formalism for the flawless propagation of EM wave through BFs structure is successfully implemented for the realistic treatment of EM wave propagation through BFs structure. Further, the theoretical formulation of optical properties of BFs structure like reflectance and transmittance at the interface of two layers, through the slab and in overall BFs structure are derived to obtain PPBG, spectra, intensity, etc. Finally, to measure and explain dispersion characteristics in existing BFs structure, various mathematical expressions are successfully derived and further discussed.

References

1. P. Yeh, A. Yariv, C.S. Hong, Electromagnetic propagation in periodic stratified media, I. General theory. J. Opt. Soc. Am. **67**, 423–437 (1977)
2. P. Yeh, A. Yariv, Bragg reflection waveguides. Opt. Commun. **19**, 427–430 (1976)
3. T. Erdogan, D.G. Hall, Circularly symmetric distribution feedback semiconductor lasers. IEEE J. Quantum Electron. **26**, 473 (1990)
4. R.K. Chourasia, V. Singh, Estimation of photonic band gap in the hollow core cylindrical multilayer structure. Superlattice. Microst. **116**, 91–199 (2018)
5. R.K. Chourasia, N.K. Chourasia, N. Bihari, *Optical Properties of Hollow-Core Bragg Fiber Waveguides*, vol 23 (Benthem Science, 2023), pp. 214–236
6. N. Bihari, A. Srivastava, N.K. Chourasia, R.K. Chourasia, H-polarised EM-wave transport in polymeric-chalcogenide columnar photonic materials. J. Phys. Conf. Ser. **2426**, 012009 (2023)
7. R.K. Chourasia, V. Singh, *Comparative analysis of the dispersion characteristic of Bragg fibers having different refractive index contrast*, in 13th International Conference on Fiber Optics and Photonics, OSA Technical Digest (online) (Optica Publishing Group, 2016), paper Th3A.75
8. R.K. Chourasia et al., Exploring optical properties in cylindrical polymeric-chalcogenides photonic materials. Mater. Today Proc. **67**(5), 625–631 (2022)
9. Y. Fink et al., Guiding optical light in air using an all-dielectric structure. J. Lightwave Technol. **17**, 2039 (1999)
10. P. Yeh, A. Yariv, E. Marom, Theory of Bragg fiber. J. Opt. Soc. Am. **68**, 1196–1201 (1978)
11. R.K. Chourasia, C.S. Yadav, A. Upadhyay, N.K. Chourasia, V. Singh, Comparative study of defect mode intensity and wavelength modulation in Bragg fiber waveguide sensors. Optik **204**, 164198 (2020)
12. R.K. Chourasia, C.S. Yadav, A. Upadhyay, N.K. Chourasia, V. Singh, Analysis of Bragg fiber waveguides having a defect layer for biosensing application. Optik **200**, 163400 (2020)
13. Y. Xu, R.K. Lee, A. Yariv, Asymptotic analysis of dielectric coaxial fibers. Opt. Lett. **27**, 1019 (2002)
14. Y. Xu et al., Asymptotic analysis of silicon based Bragg fibers. Opt. Express **11**, 1039 (2003) http://www.opticsexpress.org/abstract.cfm?URI=OPEX-11-9-1039
15. Y. Xu, R.K. Lee, A. Yariv, Asymptotic analysis of Bragg fibers. Opt. Lett. **25**, 1756 (2000)
16. Y. Xu et al., Asymptotic matrix theory of Bragg fibers. J. Lightwave Technol. **20**, 428 (2002)
17. G. Ouyang, Y. Xu, A. Yariv, Comparative study of air-core and coaxial Bragg fibers: Single mode transmission and dispersion characteristics. Opt. Express **9**, 733 (2001)
18. J.P. Meunier, J. Pigeon, J.N. Massot, A general approach to the numerical determination of modal propagation constants and field distribution of optical fibers. Opt. Quant. Electron. **13**, 71 (1981)
19. C.H. Henry, B.H. Verbeek, Solution of the scalar wave equation for arbitrarily shaped dielectric waveguides by two-dimensional Fourier analysis. J. Lightwave Technol. **7**, 308 (1989)

20. P.K. Mishra et al., Matrix method for determining propagation characteristics of optical waveguides. Opt. Acta **31**, 1041 (1984)
21. H. Etzkom, T. Heun, Highly accurate numerical method for determination of propagation dispersion flattened fibers. Opt. Quant. Electron. **18**, 1 (1986)
22. A. Sharma, S. Banarjee, Chromatic dispersion in single mode fibers with arbitrary index profiles: A simple method for exact numerical evaluation. J. Lightwave Technol. **7**, 1919 (1989)
23. D. Marcuse, Solution of the vector wave equation for general dielectric waveguides by the Galerkin method. IEEE J. Quantum Electron. **27**, 459 (1992)
24. D. Marcuse, Gaussian approximation to the fundamental modes of graded index fibers. J. Opt. Soc. Am. **68**, 103 (1978)
25. A. Weisshaar et al., Vector and quasi-vector solutions for optical waveguide modes using efficient Galerkin's method with Hermite-Gauss basis functions. J. Lightwave Technol. **13**, 1795 (1995)
26. P. Yeh, *Optical Waves in Layered Media* (Wiley, 2005)
27. N.K. Chourasia, A. Srivastava, V. Kumar, R.K. Chourasia, Doubly electrically tuned cylindrical Bragg fiber waveguide inline optical filter for multiwavelength LASER applications. Mater. Today Commun. **25**, 101620 (2020)
28. R.K. Chourasia et al., Performance study of a liquid-core Bragg fiber sensor in the presence of a defect layer. Opto-Electron. Rev. **25**(3), 215–221 (2017)
29. R.F. Cregan, B.J. Mangan, J.C. Knight, T.A. Birks, P.S. Russell, P.J. Roberts, D.C. Allan, Single-mode photonic band gap guidance of light in air. Science (1999). https://doi.org/10.1126/science.285.5433.1537
30. T. Kawanishi, M. Izutsu, Coaxial periodic optical waveguide. Opt. Express **7**, 10 (2000)
31. W. Zhi et al., Compact supercell method based on opposite parity for Bragg fibers. Opt. Express **11**, 3542–3549 (2003)
32. S.G. Johnson et al., Low-loss asymptotically single-mode propagation in large core omniguide fibers. Opt. Express **9**, 748 (2001)
33. Y. Fink et al., A dielectric omnidirectional reflector. Science **282**, 1679 (1998)
34. R.K. Chourasia, S. Prasad, V. Singh, Voltage-tunable pass band in a cylindrical multilayered structure containing PMMA and 0.67PMN–0.33PT single crystal as a defect layer. Opt. Quant. Electron. **48**, 539 (2016)
35. R.K. Chourasia, V. Singh, *Cylindrical Bragg waveguide based temperature sensors*, in 2016 IEEE Uttar Pradesh Section International Conference on Electrical, Computer and Electronics Engineering (UPCON), Varanasi, India, 2016, pp. 87–90
36. M.A. Kaliteevski, R.A. Abram, V.V. Nikolaev, G.S. Sokolovski, Bragg reflectors for cylindrical waves. J. Mod. Opt. **46**, 875–890 (1999)
37. A.H. Cherin, *An Introduction to Optical Fibers* (McGraw-Hill, Tokyo, 1983)
38. M. Born, E. Wolf, *Principles of Optics* (Cambridge University of Press, Cambridge/London, 1999)
39. S. Guo, et al., Analysis of circular fibers with arbitrary index profiles by Galerkin method. To appear in Opt. Lett. **29**, 32 (2004)
40. S. Guo, S. Albin, R.S. Rogowski, Comparative analysis of Bragg fibers. Opt. Express **12**, 198–207 (2004)
41. N.K. Chourasia, A. Srivastava, V. Kumar, R.K. Chourasia, Comparative performance study of liquid core cylindrical Bragg fibre waveguide biosensors. Pramana J. Phys. **95**, 9 (2021)
42. B. Temelkuran et al., Wavelength-scalable hollow optical fibres with large photonic bandgaps for CO2 laser transmission. Nature **420**, 650 (2002)
43. N.K. Chourasia, A. Srivastava, V. Kumar, R.K. Chourasia, Optimizing temperature-dependent molar volume fraction of biodiesel fuel in a pseudo-binary mixture through Bragg fiber waveguide sensor having defect layer. Fuel **293**, 120489 (2021)
44. M. Ibanescu et al., An all-dielectric coaxial waveguide. Science **289**, 415 (2000)
45. R.K. Chourasia, N.K. Chourasia, A. Srivastava, N. Bihari, *Photonic Nanostructured Bragg Fuel Adulteration Sensor*, vol 28 (Benthem Science, 2023), pp. 237–264

46. N.K. Chourasia, N. Bihari, R.K. Chourasia, An optofluidic Bragg fiber sensor for estimating adulterants in a temperature-dependent molar fraction of hydrated mono-alcohol fuels. Heliyon **8**(9), e10532 (2022)
47. M. Abromowitz, I.A. Stegun, *Handbook of Mathematical Functions with Formulas, Graph and Mathematical Tables* (Dover, New York, 1972)

Chapter 4
Optical Properties of Symmetrical Bragg Fiber: Periodic Structures

Abstract This chapter provides a comprehensive presentation of theoretical concepts about the propagation of electromagnetic waves in a chalcogenide (As_2Se_3)-polymeric (PEI) Bragg Fiber (CPBF). The connection between two polarization modes, E and H-polarization, is also addressed. We have presented initial numerical results for two model configurations: the single cylindrical interface and the single cylindrical slab, considering both chalcogenides (with a high refractive index) and polymeric materials (with a low refractive index). The CPBF was specifically engineered to operate within the optimal visible wavelength range of 632.8 nm. This was achieved by the utilization of advanced techniques such as the Transfer Matrix Technique (TMT) and Hankel Formalism (HF), which were applied in cylindrical coordinates. Subsequently, the characteristics of propagation and dispersion in Bragg fiber, which is composed of cladding materials with a strong refractive index contrast, are investigated for the specific wavelength range of 632.8 nm in the visible spectrum. The boundary matching approach was employed to establish a correlation between the incoming and departing light waves using TMT. A Brewster initial radius has been found where the maximum transmittance is consistently reached for both photonic materials in CPBF. The polymeric interface exhibits a lower reflectance compared to chalcogenides for the basic mode ($m = 0$). Even for higher-order modes, a rapid decrease in reflectance from 100% is observed for the polymeric contact. The reflectance response of both materials in the cylindrical unit slab exhibits both oscillating and non-oscillating components as the wavelength varies. This indicates that the initial radius of the slab significantly influences the optical characteristics, assuming a constant slab thickness. As the initial radius rises, the oscillatory reflectance becomes more squizzing. Additionally, the optical parameter exhibits an oscillating pattern in relation to the thickness of the slab, particularly for smaller mode numbers. It reaches its maximum value and remains constant for the mode number (m = 4). In addition, the observed photonic band gap (PBG) is significant in high refractive index contrast cladding BFs, and fewer periodic cladding layers are needed to obtain a perfect photonic bandgap (PPBG). The spectrum range and spectral position of PBG are significantly influenced by the angle at which the electromagnetic beam is incident. The dispersion properties have been evaluated in this chapter as well. These discoveries pave the path for many applications in the field of optoelectronic devices and sensors.

R. K. Chourasia, A. Katti, *Bragg Fibers*,
https://doi.org/10.1007/978-3-031-65164-9_4

4.1 Introduction

Researchers have been highly interested in Bragg fiber in the past decade due to its superior properties compared to standard fibers, other photonic crystal-based fibers, and Bragg gratings. The BFs were initially introduced in theory by Yeh et al. [1–3], and subsequently, researchers Doran and Blow conducted a design and feasibility analysis of cylindrical BFs for optical communications in 1983 [4]. However, by the end of the nineteenth century, a prototype for such a fiber was successfully demonstrated. BFs, or Bragg fibers, consist of a complex arrangement of concentric cylindrical cladding layers that enclose a cylindrical dielectric or hollow core with a low refractive index. The multilayer claddings are created by arranging high and low-index bilayer polymeric or chalcogenide glassy materials in a periodic pattern. These materials have good transparency in the visible region. The propagation process of electromagnetic waves (EM waves) in BFs differs significantly from that of regular fibers and waveguides that rely on total internal reflections (TIR). The layers in the BFs are intentionally constructed to have a thickness that adheres to the quarter-wave stack condition. The electromagnetic wave is propagated in optical fibers through the recurrence of Bragg reflections from the claddings. These reflections are concentrated in the low-index core of the fiber and propagate towards the opposite end. Nonlinear dispersion, distortion, core material dispersion, birefringence effects, polarization effects, and various types of losses, such as modal loss, radiation loss, and confinement loss [5–17], which are commonly observed in other traditional fibers, are effectively eliminated in BFs because of the presence of a hollow or low index core. Furthermore, the Bragg fiber's hollow or low-index core renders it very suitable for a variety of other captivating applications. Biosensors [18–28], chemical gas sensors [29], strain sensors [19, 23, 24], and other devices utilizing BFs have recently been developed. Furthermore, these unique fiber architectures have further applications such as narrowband transmission filters, optical de-multiplexers [24, 30–33], temperature sensors [34], and other optoelectronic functions.

Researchers have constructed and examined the optical properties of one-dimensional cylindrical photonic crystals. These crystals consist of a periodic arrangement of high and low refractive index materials [35–37]. They are particularly advantageous for their simple fabrication process, requiring only a single step [36]. The fabrication of BFs with high and low refractive index disparities is straightforward in this setting. Researchers have recently become interested in the potential of creating these fibers using techniques such as solvent evaporation fiber drawing and rolling [38, 39]. In the process of drawing, a preform of bilayers is initially created. Subsequently, the preform is elongated utilizing fiber drawing towers, which have been previously discussed in the second chapter of the current book. This method is used to make high-contrast BFs where the low index layer is of polymer material and the high index layer is of chalcogenide or perovskite material

[20]. However, the rolling technique is used to fabricate low-contrast BFs, i.e., both high and low index material is of polymer film, e.g., Polystyrene (PS)/Polymethyl methacrylate (PMMA) [24]. Both types of BFs can be used in sensing and other optoelectronic applications. These fibers are also naturally inspired structures [40, 41] and work well if designed at visible to near-infrared wavelength windows [42]. These working windows are important because of low loss due to hollow-core sensing and communication domains. These multilayer fibers are synthesized using the deep coating chemical root technique [43] and rolling technique to synthesize the first preform and then using fiber drawing physical technique to draw into multilayer fibers [42]. The chemical root is generally used to synthesize low R.I. contrast cladding polymeric (Poly-Ether-Sulphide, Poly-Ether-Imide, Epoxy, Poly-Metha-Methyl-Acrylate, etc.) fibers [43], while the physical root is often utilized to synthesize high R.I. contrast (more than one unit) cladding Chalcogenides-polymeric (As_2Se_3/PEI/PES, etc.) multilayer fibers [42]. From the above introduction part, one thing is clear enough the optical and other studies of such a variety of materials at the air material interface, the material's cylindrical slabs, which are the basic building blocks of such multilayer fibers are a necessity. We have thoroughly investigated their optical properties for a single interface with air and a single cylindrical slab. During this study, we have gone through the impact of initial radius, working wavelength, and slab thickness on the optical reflectance through both photonic materials.

Additionally, these BFs possess distinct characteristics that make them potentially valuable for transmitting high-power lasers, as they effectively minimize nonlinear effects [44]. Additionally, they show potential in diverse applications including strain, bio, and chemical sensing [20, 24], as well as bending and displacement sensing [38]. They can also be used for narrowband transmission filters, optical de-multiplexers, and other purposes [30, 31]. Given their recent surge in popularity, we are keen on investigating the optical characteristics of these BFs. These BFs utilize the photonic bandgap (PBG) technique to guide electromagnetic waves. Therefore, the current chapter has examined the behavior of PBG in relation to BFs. The photonic bandgap (PBG) of Bragg fibers can be adjusted by altering the angle at which light is incident. The PBG is contingent upon the spectral mode utilized during operation, such as the commonly employed visible region (for He-Ne laser at 632.8 nm) and the near-infrared regions. The band's full width at half maximum (FWHM) is contingent upon the periodicity of the BFs. Furthermore, we have also seen the photonic bandgap (PBG) shift in response to changes in the core refractive index, which serves as the fundamental principle of the biosensing process. Furthermore, in order to practically implement such BFs structure, dispersion analysis is essential. The dispersion analysis of BFs structure investigates the optimization of the core radius and cladding thickness, as well as their cutoff condition.

4.2 Theoretical and Computational Model of BFs

The TMT method is a layer-by-layer approach that provides a theoretical framework for understanding and describing the interaction of electromagnetic waves at the interface between a free medium and a polymer/chalcogenide material, as well as within the slab itself (as shown in Fig. 4.1). The Galerkin and asymptotic methodologies are incapable of assessing the photonic properties of interfaces and slabs in BFs, specifically when analyzing them slab by slab. These approaches offer less information. The TMT can accurately calculate the photonic properties of surfaces and slabs of BFs by considering factors such as the angle of incidence, thickness, Brewster's radius, planned window, and other relevant parameters. First, consider the EM field as the sum of two H-polarized opposing waves with columnar symmetry (ρ, ϕ, z). Thus H-polarized outgoing columnar wave is expressed as (ρ, ϕ, z) [45, 46]:

$$H_z^+ = AH_m^{(2)}(k)\exp(im), \tag{4.1a}$$

$$E_\phi^+ = ipAH_m^{(1)\prime}(k)\exp(im), \tag{4.1b}$$

Here "A" shows constant value, and $H_m^{(1)}$ is the first kind of Hankel function. In a similar fashion, the incoming columnar wave is expressed by the second kind Hankel function.

$$H_z^- = BH_m^{(1)}(k)\exp(im), \tag{4.2a}$$

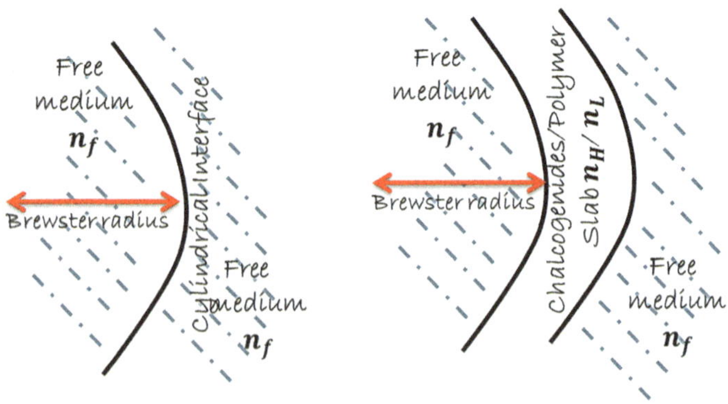

Fig. 4.1 Schematic of single cylindrical interface and single cylindrical slab for CPBFs. (Reuse of CC-BY content from Narendra Bihari, Ankita Srivastava, Nitesh K. Chourasia, and Ritesh Kumar Chourasia. "H-Polarised EM-Wave Transport in Polymeric-Chalcogenide Columnar Photonic Materials." In *Journal of Physics: Conference Series*, vol. 2426, no. 1, p. 012009. IOP Publishing, 2023)

$$E_{\phi}^{-} = ipBH_{m}^{(2)\prime}(k)\exp(im), \tag{4.2b}$$

Here, the constant value "B" likewise represents a fixed value. Therefore, the effective field component may be determined,

$$H_{z} = H_{z}^{+} + H_{z}^{-} \tag{4.3a}$$

$$E_{\phi} = ip\left[C_{m}^{(2)}H_{z}^{+} + C_{m}^{(1)}H_{z}^{-}\right] \tag{4.3b}$$

where $C_{ml}^{(1,2)} = \frac{H_{m}^{(1,2)\prime}(k)}{H_{m}^{(1,2)}(k)}$. Here "*l*" shows layer values.

The electric and magnetic field components that are tangential to the boundary between two layers exhibit continuity. The definition of a matrix $\widehat{D}$ for the transition from the basis is practical $\left(H_{z}^{+}, H_{z}^{-}\right)$ to (H_{z}, H_{ϕ}) and expressed as [45, 46]:

$$\widehat{D} = \left(1\ 1\ ipC_{m}^{(2)}\ ipC_{m}^{(1)}\right) \tag{4.4a}$$

$$\widehat{D} = \left(\frac{H_{m}^{(2)}(k)}{H_{m}^{(2)}(k\rho_{0})}\ 0\ 0\ \frac{H_{m}^{(1)}(k)}{H_{m}^{(1)}(k\rho_{0})}\right) \tag{4.4b}$$

The inverse of the above matrix is given by:

$$\widehat{D}^{-1} = \varepsilon_{r}\sqrt{\frac{\varepsilon_{0}}{\mu_{0}}}\frac{\pi}{4}KH_{m}^{(1)}(k)H_{m}^{(2)}(k)\left(-ipC_{m}^{(1)}\ 1\ ipC_{m}^{(2)}\ -1\right) \tag{4.5}$$

where $K = \frac{c}{\omega}$ is free space wavevector, while the symbol "*p*" represents the intrinsic impedance of the medium. The requirement for continuity of the tangential field components at the intersection of layers "1" and "2" can be described as follows,

$$\left(H_{z2}^{+}\ H_{z2}^{-}\right) = \widehat{D_{2}}^{-1}\widehat{D_{1}}\left(H_{z1}^{+}\ H_{z1}^{-}\right) \tag{4.6}$$

$$D_{11} = -i\varepsilon_{r}\sqrt{\frac{\varepsilon_{0}}{\mu_{0}}}\frac{\pi}{4}KH_{m}^{(1)}(k_{2})H_{m}^{(2)}(k_{2})\left[p_{2}C_{m2}^{(1)} - p_{1}C_{m1}^{(2)}\right], \tag{4.7a}$$

$$D_{21} = -i\varepsilon_{r}\sqrt{\frac{\varepsilon_{0}}{\mu_{0}}}\frac{\pi}{4}KH_{m}^{(1)}(k_{2})H_{m}^{(2)}(k_{2})\left[p_{1}C_{m1}^{(2)} - p_{2}C_{m2}^{(2)}\right], \tag{4.7b}$$

$$D_{12} = -i\varepsilon_{r}\sqrt{\frac{\varepsilon_{0}}{\mu_{0}}}\frac{\pi}{4}KH_{m}^{(1)}(k_{2})H_{m}^{(2)}(k_{2})\left[p_{2}C_{m2}^{(1)} - p_{1}C_{m1}^{(1)}\right], \tag{4.7c}$$

$$D_{22} = -i\varepsilon_r\sqrt{\frac{\varepsilon_0}{\mu_0}}\frac{\pi}{4}KH_m^{(1)}(k_2)H_m^{(2)}\left(k_2\right)\left[p_1C_{m1}^{(1)} - p_2C_{m2}^{(2)}\right]. \tag{4.7d}$$

4.2.1 Photonic Properties at Columnar Interfaces

To determine the reflectance at the columnar interface, we need to apply the transfer matrix $\widehat{T} = \widehat{D_2}^{-1}\widehat{D_1}$ to the incident electromagnetic wave on the two-layer interface.

$$(t_d\ 0\ ) = \widehat{T}(1\ r_d\ ) \tag{4.8}$$

Form Eqs. (4.7) and (4.8), we have [45, 6],

$$r_d = \frac{p_2C_{m2}^{(2)} - p_1C_{m1}^{(2)}}{p_1C_{m1}^{(1)} - p_2C_{m2}^{(2)}} \tag{4.9}$$

4.2.2 Photonic Properties in Columnar Slab

Additionally, the reflectance value can also be determined in a columnar slab. The application of Hankel formalism and TM approach is applicable to the columnar slab depicted in Fig. 4.1. The reflectance coefficient for a columnar slab may be expressed explicitly as [45, 46],

$$r_d - r_{0d} = Cr_{1d}\frac{H_m^{(2)}(k_1\rho_1)}{H_m^{(1)}(k_1\rho_1)}t_{0c}H_m^{(1)}(k_1\rho_0) \tag{4.10}$$

The optical reflectance is determined by this equation,

$$R = |r_d|^2 \tag{4.11}$$

Therefore, in the case of nonabsorbing media, the equation provides the transmittance across the interface and slab medium,

$$T = 1 - R \tag{4.12}$$

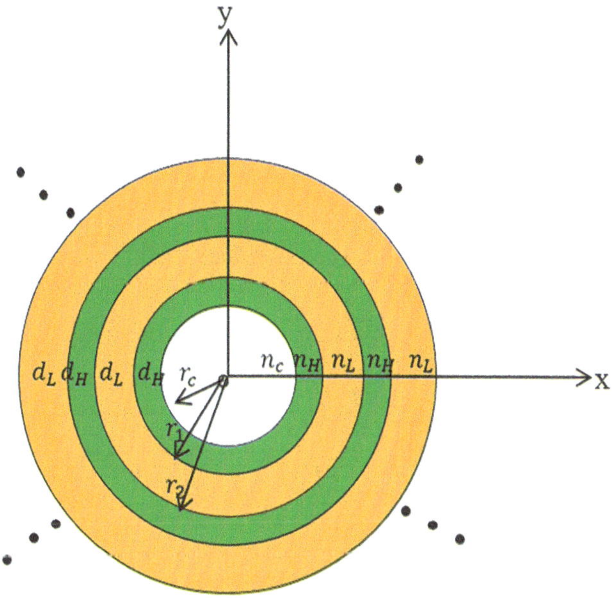

Fig. 4.2 Schematic layer-by-layer representation of BFs showing layers' position, thickness and refractive index as well propagation of EM wave in z-direction

Therefore, Eqs. (4.9), (4.10), (4.11), and (4.12) are highly valuable for determining the transmittance of the interface and columnar slab polymer/chalcogenides materials.

4.2.3 Photonic Properties in Overall BFs

Figure 4.2 illustrates a cross-sectional view of the BFs along their length. The figure displays the refractive index (R.I.) and radius of the core as n_c, r_c, while n_H, n_L and d_H, d_L represent the RIs and thicknesses of the high and low index layers, respectively. These layers combine to form a bilayer structure. $\Lambda = d_H + d_L$ is the bilayer period thickness.

Let's assume that the field components have a spatial-temporal factor of $U_{tz} = \exp[i(\omega t - \beta z)]$, where β represents the propagation constant. By repeating these layers, it is possible to construct a multilayer structure known as bilayer films.

Within the context of BFs, the electromagnetic wave is expected to possess a cylindrical shape as a result of the cylindrical symmetry inherent in BFs. The cylindrical wave is presumed to be radiating away from the axis of symmetry $r = 0$, and subsequently encounters the initial cylindrical contact $r = r_c$. Now, utilizing the TMT method for the cylindrical waves [45] to compute the reflectance of the BFs: The temporal component of all the fields is considered to be exp($i\omega t$). Therefore, the Maxwell's equations without sources [47] can be expressed as,

$$\nabla \times E = -j\omega H \tag{4.13a}$$

$$\nabla \times H = j\omega E \tag{4.13b}$$

The nonzero fields E_z, H_θ, H_r in each layer fulfill the following three equations for TE mode or H polarization in cylindrical coordinates (r, θ, z) [47],

$$\frac{1}{r}\frac{\partial E_z}{\partial \theta} = -j\omega\mu H_r \tag{4.14a}$$

$$\frac{\partial E_z}{\partial r} = j\omega\mu H_\theta \tag{4.14b}$$

$$\frac{\partial (rH_\theta)}{\partial r} - \frac{\partial H_r}{\partial \theta} = -j\omega\varepsilon r E_z. \tag{4.14c}$$

The governing differential equation for the tangential component of the electric field E_z can be expressed by utilizing Eq. (4.14c),

$$r\frac{\partial}{\partial r}\left(r\frac{\partial E_z}{\partial r}\right) - r^2\frac{1}{\mu}\frac{\partial \mu}{\partial r}\frac{\partial E_z}{\partial r} + \frac{\partial}{\partial \theta}\left(\frac{\partial E_z}{\partial \theta}\right) + \omega^2\varepsilon\mu r^2 E_z = 0. \tag{4.15}$$

Now, let us assume the field component $E_z = U(r)\Theta(\theta) = U(r)e^{im\theta}$, and the solution U is given as;

$$U(r) = C_1 J_m(kr) + C_2 Y_m(kr), \tag{4.16}$$

where C_1 and C_2 are constants, J_m is a Bessel function, Y_m is the Neumann function, and k is the wave number in the material medium. In this investigation, the azimuthal mode number, denoted by the subscript "m," has been set to zero. Similarly, the magnetic field's tangential component can be represented as, $H_\theta = V(r)e^{im\theta}$, where $V(r)$ is given by,

$$V(r) = -jp\left(C_1 J'_m(kr) + C_2 Y'_m(kr)\right), \tag{4.17}$$

where the material medium's intrinsic admittance is denoted by $p = \sqrt{\varepsilon/\mu}$.

The construction of a monolayer matrix that establishes a relationship between the electric and magnetic fields at its two interfaces is achieved by simultaneously employing Eqs. (4.16) and (4.17). For instance, the matrix $\widehat{T}_1$ for the first layer at the interface ($r = r_0$ and r_1) is written as [14, 47]:

$$\begin{bmatrix} U(r_1) \\ V(r_1) \end{bmatrix} = \widehat{T}_1 \begin{bmatrix} U(r_0) \\ V(r_0) \end{bmatrix}. \tag{4.18}$$

The component of the single-layer matrix $\widehat{T}_1 = \begin{bmatrix} t_{11} & t_{12} \\ t_{21} & t_{22} \end{bmatrix}$ can be written as

$$t_{11} = \frac{\pi}{2} k_1 r_0 \left[Y'_m(k_1 r_0) J_m(k_1 r_1) - J'_m(k_1 r_0) Y_m(k_1 r_1) \right], \tag{4.19a}$$

$$t_{12} = j\frac{\pi}{2} \frac{k_1}{p_1} r_0 \left[J_m(k_1 r_0) Y_m(k_1 r_1) - Y_m(k_1 r_0) J_m(k_1 r_1) \right], \tag{4.19b}$$

$$t_{21} = -j\frac{\pi}{2} k_1 r_0 p_1 \left[Y'_m(k_1 r_0) J'_m(k_1 r_1) - J'_m(k_1 r_0) Y'_m(k_1 r_1) \right], \tag{4.19c}$$

$$t_{22} = \frac{\pi}{2} k_1 r_0 \left[J_m(k_1 r_0) Y'_m(k_1 r_1) - Y_m(k_1 r_0) J'_m(k_1 r_1) \right], \tag{4.19d}$$

where $p_1 = \sqrt{\varepsilon_1/\mu_1}$. Equation (4.19) demonstrates that the matrix elements are solely determined by the radii of the two interfaces. Therefore, one may easily create the matrix for any existing layer of BFs. To generate the total transfer matrix $\ddot{T}$ that relates the first and final interfaces, a total of 2 N matrices are required for the BFs.

$$\begin{bmatrix} U(r_f) \\ V(r_f) \end{bmatrix} = \widehat{T} \begin{bmatrix} U(r_0) \\ V(r_0) \end{bmatrix}, \tag{4.20}$$

where

$$\widehat{T} = \begin{bmatrix} T_{11} & T_{12} \\ T_{21} & T_{22} \end{bmatrix} = \widehat{T}_{2\mathrm{N}} \widehat{T}_{2\mathrm{N}-1} \ldots \ldots \widehat{T}_{\mathrm{N}+1} \widehat{T}_{\mathrm{N}} \ldots \ldots \widehat{T}_2 \widehat{T}_1. \tag{4.21}$$

The reflection and transmission coefficients can be obtained using the computed transfer matrix by using the following relationships [45],

$$r_d = \frac{\left(T'_{21} + jp_0 C^{(2)}_{m0} T'_{11}\right) - jp_f C^{(2)}_{mf}\left(T'_{22} + jp_0 C^{(2)}_{m0} T'_{12}\right)}{\left(-jp_0 C^{(1)}_{m0} T'_{11} - T'_{21}\right) - jp_f C^{(2)}_{mf}\left(-jp_0 C^{(1)}_{m0} T'_{12} - T'_{22}\right)}, \tag{4.22}$$

$$t_d = \frac{4\sqrt{\varepsilon_0/\mu_0}}{\pi K r_0 H^{(2)}_m(k_0 r_0) H^{(1)}_m(k_0 r_0)\left[\left(-jp_0 C^{(1)}_{m0} T'_{11} - T'_{21}\right) - jp_f C^{(2)}_{mf}\left(-jp_0 C^{(1)}_{m0} T'_{12} - T'_{22}\right)\right]}, \tag{4.23}$$

where $p_0 = \sqrt{\varepsilon_0/\mu_0}$ and $p_\mathrm{f} = \sqrt{\varepsilon_\mathrm{f}/\mu_\mathrm{f}}$ are the admittance of core and final media, $T'_{11}, T'_{12}, T'_{21}$ and T'_{22} are the matrix element of the inverse matrix of $\hat{T}$, $K = \omega\sqrt{\varepsilon_0\mu_0}$ is the free-space wavenumber and

$$C^{(1,2)}_{ml} = \frac{H^{(1,2)\prime}_m(k_1 r_1)}{H^{(1,2)}_m(k_1 r_1)}, l = 0, f. \tag{4.24}$$

where $H^{(1)}_m$ and $H^{(2)}_m$ are the Hankel function of the first and second kind. By utilizing Eqs. (4.22) and (4.23), the reflectance and transmittance percentage can be determined,

$$\%R = 100 \times |r_d|^2, \%T = 100 \times \frac{n_\mathrm{out}}{n_\mathrm{c}} |t_d|^2. \tag{4.25}$$

The refractive indices of the core and outer media are denoted as n_c and n_out, respectively.

4.2.4 Hankel Formulation in BFs (for Dispersion Characteristics)

Figure 4.3 displays the schematic RI profile of the BFs. This proposed BFs structure adheres to both TE and TM modes. However, in this particular chapter, our attention

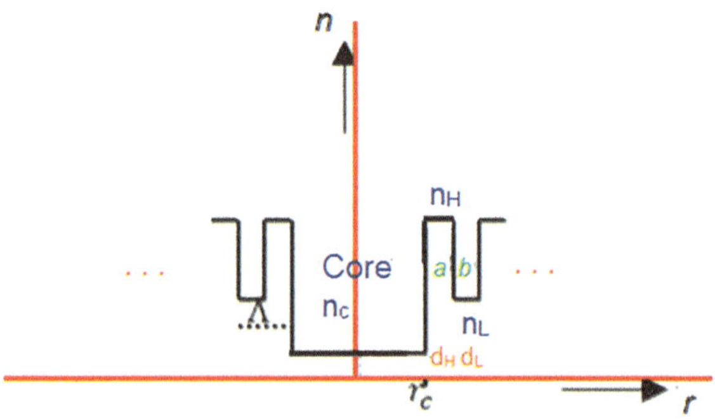

Fig. 4.3 Schematic R.I. profile of BFs structure along with the layers' radial positions

has been solely directed towards the TE mode, which exhibits field components H_r, E_θ and H_z that are not equal to zero. Subsequent components may be derived from the aforementioned ones. It is chosen that the fundamental components of electromagnetic fields remain continuous at the interface of every layer. Accordingly, H_z, E_θ are therefore selected as the fundamental components in TE mode for the subsequent calculation. Additionally, the EM fields mentioned earlier for TE mode can be represented as matrices [14];

$$(H_z \; iE_\theta) = U_{tz} D_i^{TE}(r)(A_i \; B_i) \tag{4.26}$$

where, $D_i^{TE}(r) \equiv \left(d_{11}^{TE} \; d_{12}^{TE} \; d_{21}^{TE} \; d_{22}^{TE}\right)$ is referred to as the representation matrix. Each element of the representation matrix in the core region is defined as follows,

$$d_{11}^{TE} = H_\nu^{(2)}(\kappa_c r), \tag{4.27a}$$

$$d_{12}^{TE} = H_\nu^{(1)}(\kappa_c r), \tag{4.27b}$$

$$d_{21}^{TE} = -\left(\frac{\omega\mu_0}{\kappa_c}\right) H_\nu^{(2)'}(\kappa_c r), \tag{4.27c}$$

$$d_{22}^{TE} = \left(\frac{n_c^2 \omega\varepsilon_0}{\kappa_c}\right) H_\nu^{(1)'}(\kappa_c r). \tag{4.27d}$$

$$\kappa_c \equiv \sqrt{(n_c k_0)^2 - \beta^2} \; ; 0 \le r \le r_c. \tag{4.28}$$

Hankel functions, $H_\nu^{(1)} = J_\nu + iN_\nu$ and $H_\nu^{(2)} = J_\nu - iN_\nu$, represent the inward and external waves, respectively, whereas the first differentiation of the argument is denoted by prime. Further, the wavenumber in vacuum is $k_0 = \frac{\omega}{c}$. In order to represent fields in claddings for the BFs structure, which has a relatively large core radius, an asymptotic expansion for the Hankel function at large arguments [48] can be utilized. The representation matrix is illustrated in this instance by means of the boundary matrix P_i^{TE} along with Q_i^{TE}, the displacement matrix as,

$$D_i^{TE}(r) \simeq \sqrt{\frac{2}{\pi\rho}} P_i^{TE} Q_i(r); \tag{4.29}$$

where

$$P_i^{TE} = \frac{1}{\sqrt{\kappa_i}} \left(1 \; 1 \; i\omega\mu_0/\kappa_i - i\omega\mu_0/\kappa_i\right) \tag{4.30a}$$

$$Q_i^{TE}(\rho) = (\exp(-i\kappa_i r)\ 0\ 0\ \exp(i\kappa_i r)\ ); \tag{4.30b}$$

The subscript i denotes layers a, b, and c, as illustrated in Fig. 4.2. In TE mode, the amplitude pair for layer "a" is correlated,

$$(a_{m+1}\ b_{m+1}\ ) = (X_{TE}\ Y_{TE}\ Y_{TE}^*\ X_{TE}^*\ )(a_m\ b_m\ ); \tag{4.31a}$$

$$X_{TE} = \left[\text{coscos}\ (\kappa_b d_L) - \frac{i}{2}\left(\frac{\kappa_b}{\kappa_a} + \frac{\kappa_a}{\kappa_b}\right)\text{sinsin}\ (\kappa_b d_L)\ \right] \exp(-i\kappa_a d_H); \tag{4.31b}$$

$$Y_{TE} = \frac{i}{2}\left(\frac{\kappa_b}{\kappa_a} - \frac{\kappa_a}{\kappa_b}\right)\text{sinsin}\ (\kappa_b d_L)\ ]\exp(i\kappa_a d_H) \tag{4.31c}$$

The approximate expression $|X_{TE}|^2 - |Y_{TE}|^2 = 1$ in the the asymptotic limit is satisfied by each coefficient. By applying the Floquet theorem [1–3, 48] to the periodic structure, the eigenvalue equation for the BFs structure can be determined. As stated by Floquet, in the case where a substance possess the periodicity, $n(r + \Lambda) = n(r)$, the wave that propagates through that medium can be described as follows:

$$G_K(r, z) = G_K(r)\exp(-iKr)\exp(-i\beta z) \tag{4.32}$$

In this context, K represents the Bloch wavenumber, which is a universal value in all periodic structures. The eigenvalue equation can be obtained by analyzing the boundary condition at the contact between the core and cladding. It yields

$$\frac{J_\nu'(\kappa_c r_c)}{J_\nu(\kappa_c r_c)} + \frac{\kappa_c}{i\kappa_a}\frac{\{[\exp\ (-iK^{TE}\Lambda) - X_{TE}] - Y_{TE}\}}{\{[\exp\ (-iK^{TE}\Lambda) - X_{TE}] + Y_{TE}\}} = 0 \tag{4.33}$$

Multiple attributes can be assessed from the aforementioned equation.

4.3 Obtained Outcomes and Discussion

4.3.1 Columnar Interface and Slab Analysis

In order to optimize and investigate the optical properties, two materials with a significant difference in refractive index have been selected: the polymeric material Poly-Ether-Imide (PEI) with a refractive index of $n_L = 1.66$ [42], and the chalcogenides material As_2Se_3 with a refractive index of $n_H = 2.82$ [43]. The selected materials have been created using a window-designed wavelength (632.8 nm) with minimal loss. The materials are optimized to have a width that corresponds to a quarter of the wavelength. The primary objective of our current study activity is to

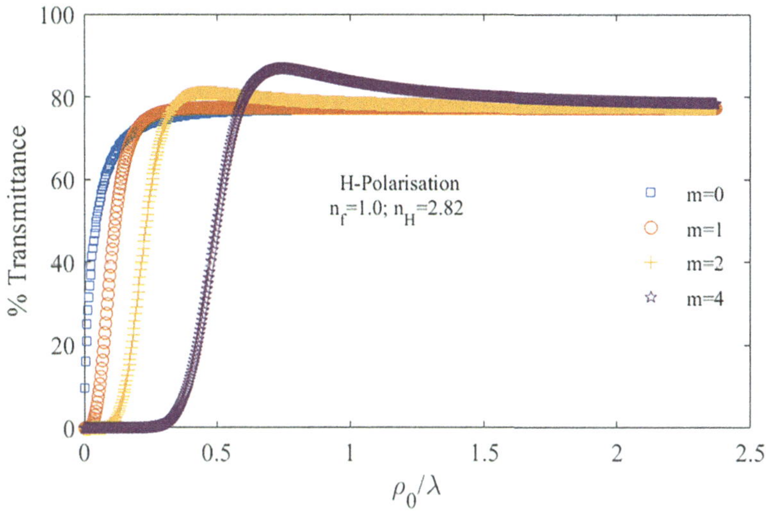

Fig. 4.4a Transmittance vs. standardized radius plot at empty core/chalcogenides junction for different "*m*" values. (Reuse of CC-BY content from Narendra Bihari, Ankita Srivastava, Nitesh K. Chourasia, and Ritesh Kumar Chourasia. "H-Polarised EM-Wave Transport in Polymeric-Chalcogenide Columnar Photonic Materials." In *Journal of Physics: Conference Series*, vol. 2426, no. 1, p. 012009. IOP Publishing, 2023)

investigate the columnar empty medium/material boundary and the columnar slab of photonic materials, as depicted in Fig. 4.1. The initial focus is on the transmittance at the interface between two regions of unobstructed medium and photonic materials. Unlike the usual standard interface, the initial radius (known as Brewster's radius) must also be accounted for when analyzing the propagation of electromagnetic waves in columnar systems. Figures 4.4a and 4.4b display graphs illustrating the relationship between transmittance and scaled radius for various H-polarized azimuthal modes at the boundary between the empty medium and polymeric/chalcogenides materials. The data indicate that the transmittance at ($m = 0$) is lower for both materials compared to higher ($m > 0$) values. However, the transmittance measured for chalcogenide materials is substantially lower. Additionally, for lower azimuthal mode $m = 0$, the transmittance increases until ($\frac{\rho_0}{\lambda} < 0.3$), while for higher azimuthal modes, the transmittance increases until ($\frac{\rho_0}{\lambda} < 0.6$) for both materials, remaining steady thereafter. An increase in transmittance is observed with positive values of m. Initial transmittance is 0, escalates rapidly with increasing radius value, and ultimately converges at a significantly large radius value. The Brewster's phenomena results in a sudden increase for $m > 0$ at a specific radius, which shifts towards larger values as the "m" values increase. Therefore, that particular radius is referred to as Brewster's radius.

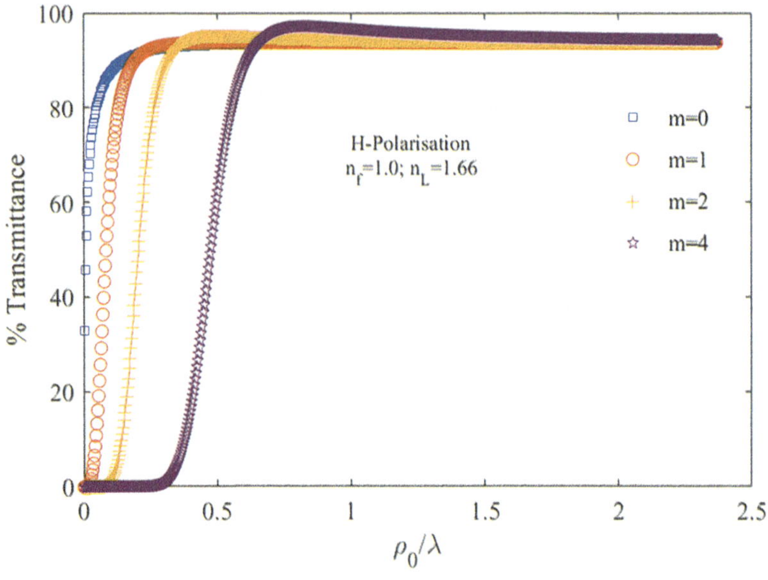

Fig. 4.4b Transmittance vs. standardized radius plot at empty core/polymeric material junction for different "*m*" values. (Reuse of CC-BY content from Narendra Bihari, Ankita Srivastava, Nitesh K. Chourasia, and Ritesh Kumar Chourasia. "H-Polarised EM-Wave Transport in Polymeric-Chalcogenide Columnar Photonic Materials." In *Journal of Physics: Conference Series*, vol. 2426, no. 1, p. 012009. IOP Publishing, 2023)

Both materials are supposed to have a columnar slab shape and are submerged in a media that is not restricted. The transmittance vs. standardized wavelength graphs for the polymeric and chalcogenides columnar slabs are shown in Figs. 4.5a, 4.5b, 4.5c, and 4.5d. These plots depict the relationship between transmittance and wavelength for two different starting radius values: $\rho_0 = 100$ nm and $\rho_0 = 1000$ nm. The four graphs display both the oscillating and non-oscillating components of the transmittance, which vary with the wavelength. Both chalcogenides and polymeric materials exhibit higher transmittance in the oscillatory region, whereas it decreases in the non-oscillatory zone for certain "*m*" values at a smaller initial radius. However, this tendency is reversed for lower azimuthal modes for both materials. As the initial radius increases rapidly, the oscillatory component becomes compressed in both materials. The oscillatory behaviors also exhibit a considerable dependence on the initial radius. As the initial radius and photonic bandgap (PBG) increase, it signifies that the range of frequencies in which electromagnetic (EM) waves may propagate becomes increasingly limited. In each instance, when the value of "*m*" is bigger (specifically, $m = 4$), absolute null transmittance is seen.

Lastly, we will investigate the transmittance that is dependent on the breadth. By manipulating the design wavelengths while keeping the initial radius fixed, it is possible to adjust the width. Figures 4.6a and 4.6b depict the relationship between transmittance and slab thickness for different materials. The data unambiguously

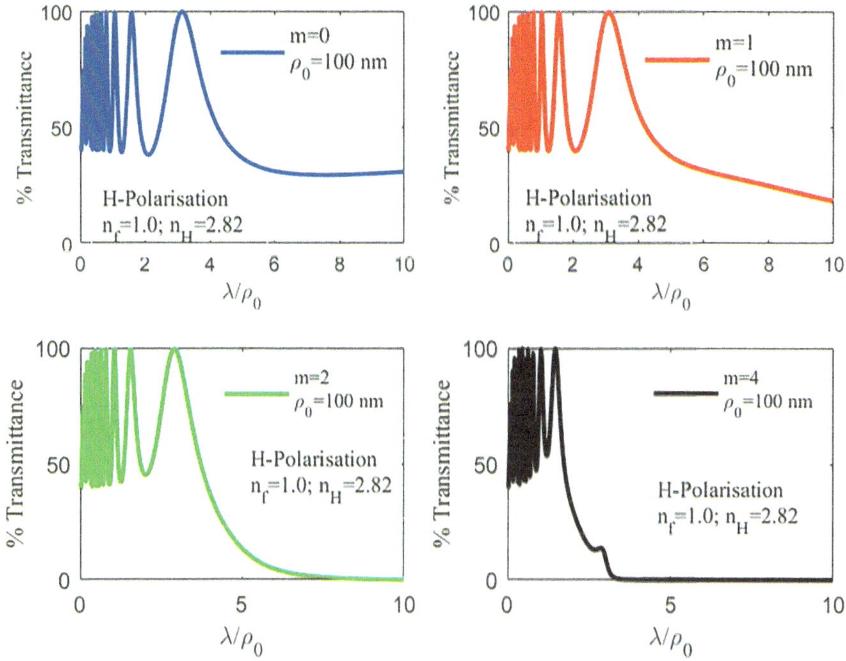

Fig. 4.5a Transmittance vs. standardized wavelength (at starting radius $\rho_0 = 100$ nm) plot of chalcogenides columnar slab at various "*m*" values. (Figs. 4.5a–d are Reuse of CC-BY content from Narendra Bihari, Ankita Srivastava, Nitesh K. Chourasia, and Ritesh Kumar Chourasia. "H-Polarised EM-Wave Transport in Polymeric-Chalcogenide Columnar Photonic Materials." In *Journal of Physics: Conference Series*, vol. 2426, no. 1, p. 012009. IOP Publishing, 2023)

demonstrate that columnar materials exhibit distinct behavior compared to traditional photonic materials, since the transmittance increases proportionally with the width of the slab. Conversely, the transmittance in our present circumstances varies periodically with the breadth of the slab. Chalcogenides materials exhibit a greater magnitude of oscillatory transmittance value when compared to polymeric materials. Furthermore, the transmittance of both materials with varying slab width declines as the "*m*" values change and reaches zero when m equals 4.

4.3.2 The Overall BFs Structure Analysis

The study focuses on the propagation characteristics of high refractive index contrast Bragg fibers in the visible area, specifically at a design wavelength of 632.8 nm. The fiber core and outer layer have to be taken air, i.e., $n_c = n_{\text{out}} = 1.0$ with core dimension small enough $r_c = 100(d_H + d_L)$ to avoid the losses that occur due to the presence of higher-order modes. The high R.I. contrast BFs is composed of high

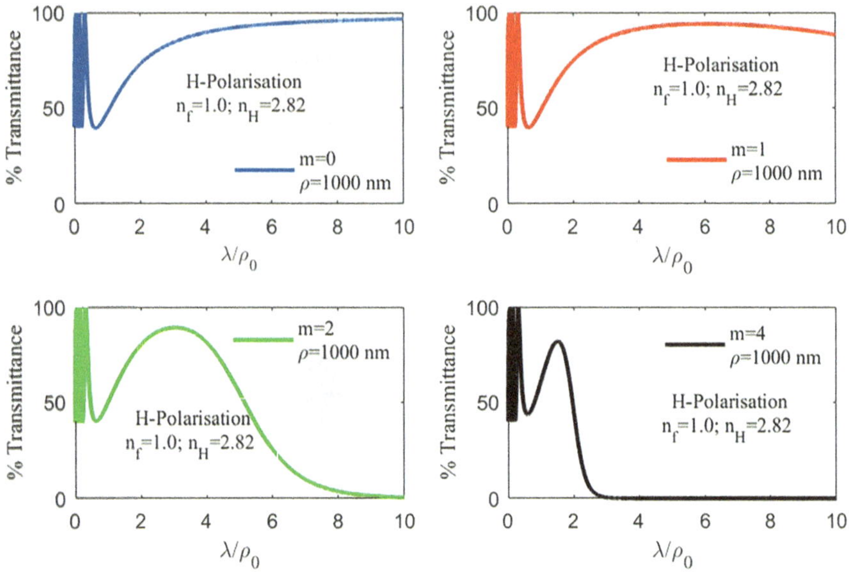

Fig. 4.5b Transmittance vs. standardized wavelength (at starting radius $\rangle_0 = 1000$ nm) plot of chalcogenides columnar slab at various "*m*" values. (Figs. 4.5a–d are Reuse of CC-BY content from Narendra Bihari, Ankita Srivastava, Nitesh K. Chourasia, and Ritesh Kumar Chourasia. "H-Polarised EM-Wave Transport in Polymeric-Chalcogenide Columnar Photonic Materials." In *Journal of Physics: Conference Series*, vol. 2426, no. 1, p. 012009. IOP Publishing, 2023)

R.I. chalcogenide material (As_2Se_3), n_H=2.82 and the low R.I. polymer material (PEI) n_L=1.66 [18] having a respective thickness $d_H = \lambda_c/4n_H$ and $d_L = \lambda_c/4n_L$ [49] that follows the quarter-wave stack condition.

Figure 4.7 displays the relationship between the percentage of light transmittance and the wavelength of incident light for high refractive index (R.I.) contrast brightfield (BF) samples at the specific design wavelength of 632.8 nm. The perfect photonic bandgap (PPBG) is a certain range of wavelengths where there is no transmission of light, resulting in 0% transmittance. The measured PPBG for this specific example is approximately 225 nm. The width fluctuation of the photonic band gap (PBG) and the corresponding percentage transmittance are calculated and presented in Table 4.1 for different numbers of unit cells (N) in this specific BF.

In order to observe the impact of different angles at which electromagnetic waves strike the photonic bandgap structure (PPBG), a total of 14 unit cells are selected in high refractive index contrast Bragg filters (BFs) for the specific design wavelength of 632.8 nm. Figure 4.8 displays the spectrum of photonic band gap (PBG) at various angles of incidence. The positions of the left band edge (LBE) and right band edge (RBE) for this specific BFs are determined from the given figure and recorded in Table 4.2. The table clearly demonstrates that with High R.I. contrast BFs, a change in angle of incidence from 00 to 750 can cause a leftward movement of the band edge by 64 nm and a rightward shift of the band edge by approximately 94 nm. The

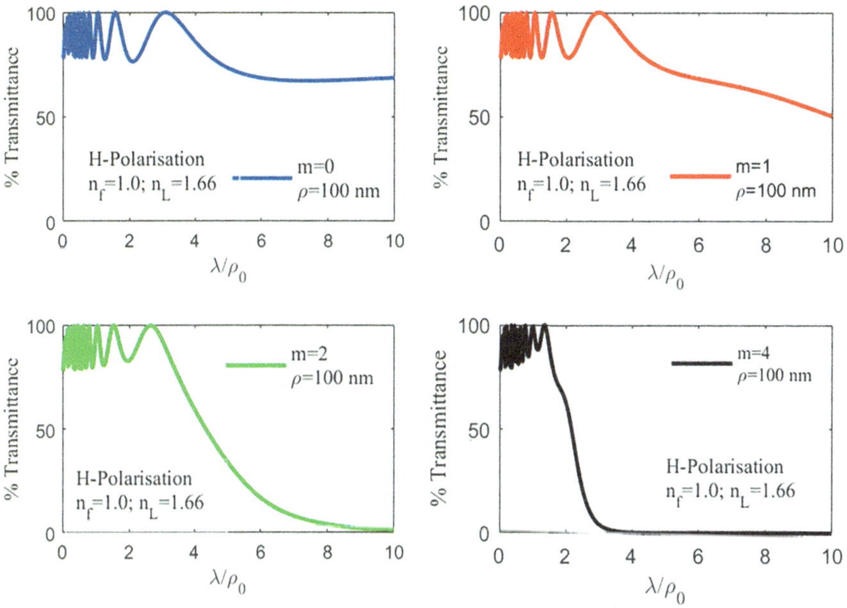

Fig. 4.5c Transmittance vs. standardized wavelength (at starting radius $\rangle_0 = 100$ nm) plot of polymeric material columnar slab at various "*m*" values. (Figs. 4.5a–d are Reuse of CC-BY content from Narendra Bihari, Ankita Srivastava, Nitesh K. Chourasia, and Ritesh Kumar Chourasia. "H Polarised EM-Wave Transport in Polymeric-Chalcogenide Columnar Photonic Materials." In *Journal of Physics: Conference Series*, vol. 2426, no. 1, p. 012009. IOP Publishing, 2023)

apparent reduction in wavelength with increasing incidence angles is a result of the light ray's geometric path lengthening. In addition, the Photonic Band Gaps (PBGs) experience a blue shift as the geometrical path of the incident light beam is increased.

The utilization of PBG as a sensing signal allows for the application of high refractive index contrast BFs structure in chemical and biosensing applications. Fig. 4.9 illustrates the changes in the positions of LBE, RBE, and PBG with variations in the refractive indices of the core in high refractive index contrast Bragg fibers operating at a wavelength of 632.8 nm. These changes are also summarized in Table 4.3 to assess the potential use of these fibers in biosensing applications. Table 4.3 indicates that, for sensing purposes, one can observe either the variation in Full Width at Half Maximum (FWHM) of Photonic Bandgap (PBG), or the shift in the spectral position of PBG. Upon analyzing the spectral shift in the hypothesized BFs, it is evident that the RBE exhibits a significantly greater magnitude than the LBE. Hence, it is recommended that monitoring RBE instead of LBE will yield superior outcomes.

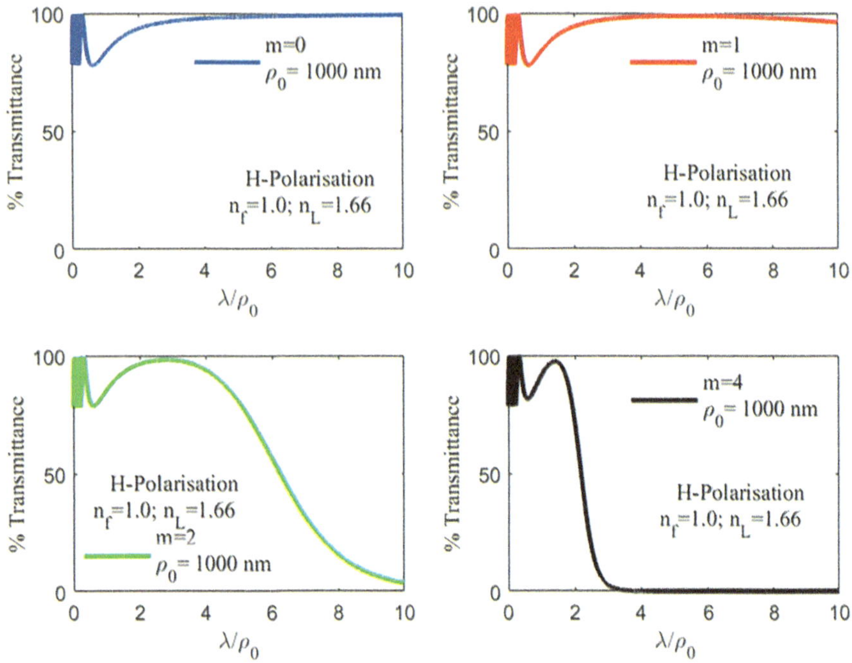

Fig. 4.5d Transmittance vs. standardized wavelength (at starting radius $\rangle_0 = 1000$ nm) plot of polymeric material columnar slab at various "*m*" values. (Figs. 4.5a–d are Reuse of CC-BY content from Narendra Bihari, Ankita Srivastava, Nitesh K. Chourasia, and Ritesh Kumar Chourasia. "H-Polarised EM-Wave Transport in Polymeric-Chalcogenide Columnar Photonic Materials." In *Journal of Physics: Conference Series*, vol. 2426, no. 1, p. 012009. IOP Publishing, 2023)

In contrast, in order to examine the dispersion properties, the hollow core of the BFs has to be selected, i.e., $n_c = 1.0$ and for high R.I. contrast BFs, high R.I. cladding is (As_2Se_3), $n_H = 2.82$ and low R.I. cladding is polyether-imide (PEI), $n_L = 1.66$ [18]. The thicknesses of the cladding layers d_H and d_L on both waveguides conform to the quarter-wave stack condition, $n_H d_H = n_L d_L = \frac{\lambda_0}{4}$, where, λ_0 is the design wavelength. A design wavelength of 632.8 nm has been selected for the He-Ne laser pulse in order to facilitate communication and sensing. The normalized propagation constant β/k_0 for several TE modes including the fundamental mode can be plotted as a function of core radius r_c at 632.8 nm [21]. The cutoff radius r_c for various $TE_{0\mu}$ modes is tabulated above which shows that the cutoff radius for fundamental mode is lower at design wavelength 632.8 nm. More no of modes are present in high R.I. contrast (Table 4.4).

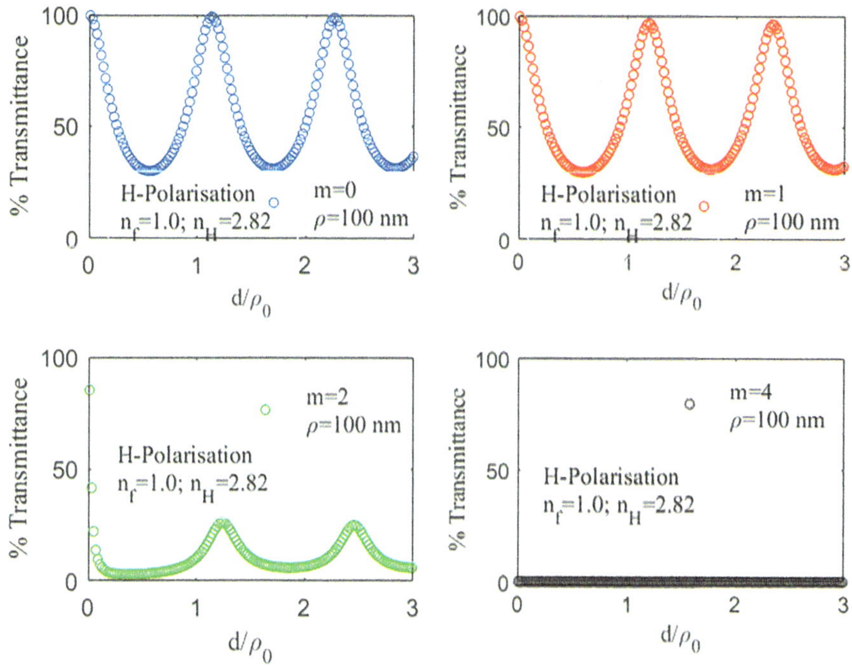

Fig. 4.6a Transmittance vs. standardized chalcogenide slab width at different "*m*" values. (Figures 4.6a, b are Reuse of CC-BY content from: Narendra Bihari, Ankita Srivastava, Nitesh K. Chourasia, and Ritesh Kumar Chourasia. "H-Polarised EM-Wave Transport in Polymeric-Chalcogenide Columnar Photonic Materials." In *Journal of Physics: Conference Series*, vol. 2426, no. 1, p. 012009. IOP Publishing, 2023)

4.4 Conclusion

By observing the propagation of h-polarized electromagnetic waves within the interface and cylindrical portion of two distinct varieties of chalcogenide-polymeric photonic materials, the fundamental optical properties of these materials were examined. The obtained optical transmittance at cylindrical interfaces and slabs is confirmed to be strongly dependent on the H-polarized azimuthal modes, initial radius, and slab thicknesses by the theoretical formulation. A critical value of Brewster's radius exists for maximum optical transmittance, which rises in proportion to the azimuthal mode number. Furthermore, the wavelength-dependent optical transmittance of the cylindrical slab can be categorized into oscillatory and non-oscillatory regions. These regions have a significant impact on the transmission window and PBG values. Intriguing oscillatory behaviors are also observed in the slab thickness-transmittance plots, which are in stark contrast to the behavior observed in planar photonic materials. Additionally, an investigation is conducted into the propagation characteristics of hollow-core high R.I. contrast BFs within a wavelength window of interest. Only a limited number of periodicities are required

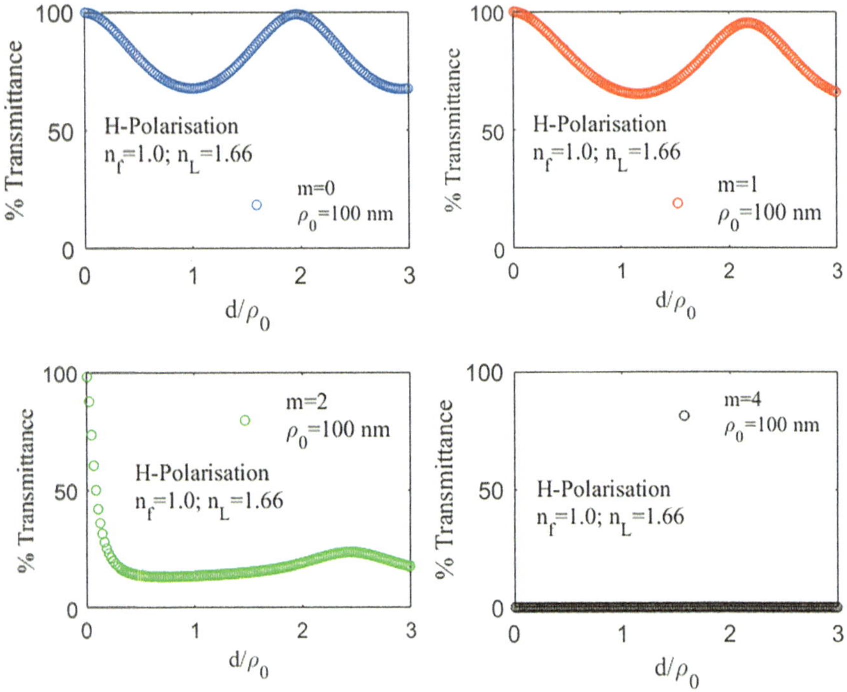

Fig. 4.6b Transmittances vs. standardized polymeric material slab width at different "*m*" values. (Figures 4.6a, b are from: Narendra Bihari, Ankita Srivastava, Nitesh K. Chourasia, and Ritesh Kumar Chourasia. "H-Polarised EM-Wave Transport in Polymeric-Chalcogenide Columnar Photonic Materials." In *Journal of Physics: Conference Series*, vol. 2426, no. 1, p. 012009. IOP Publishing, 2023)

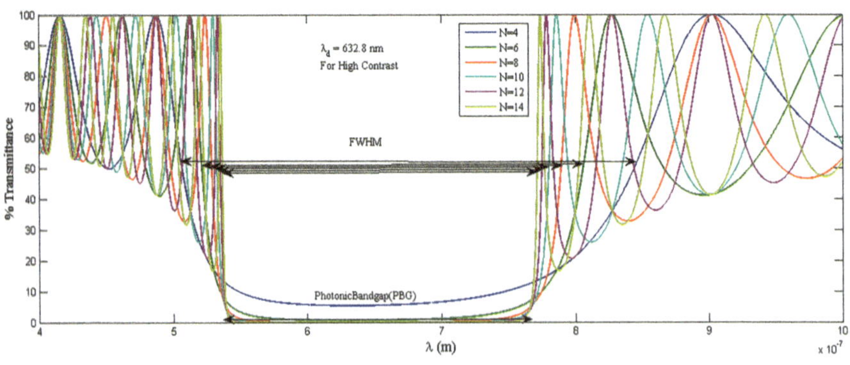

Fig. 4.7 Different cladding layers impact on PBG at design wavelengths of 632.8 nm for high R.I. contrast BFs. (Reprinted from [25]: Superlattices and Microstructures, 116, Ritesh Kumar Chourasia, Vivek Singh, Estimation of photonic band gap in the hollow core cylindrical multilayer structure, 191–199, Copyright 2018, with permission from Elsevier)

Table 4.1 Variability in FWHM and percentage transmittance at design wavelengths of 632.8 nm for high R.I. contrast BFs [25]

S.no.	High R.I. contrast BFs at design wavelength 632.8 nm						
1.	N	4	6	8	10	12	14
2.	%T	5.4520	0.6642	0.0793	0.0094	0.0011	0.0001
3.	FWHM (nm)	341.0	285.0	263.2	246.1	235.0	210.0

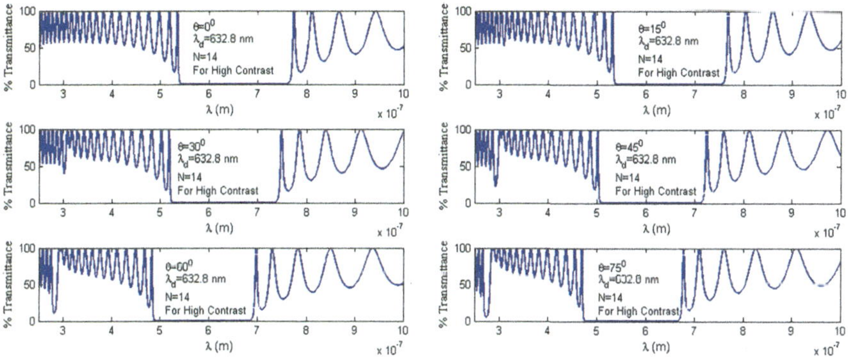

Fig. 4.8 Spectral shift of obtained PBG with the angle of incidence high R.I. contrast BFs for design wavelength 632.8 nm. (Reprinted from [25]: Superlattices and Microstructures, 116, Ritesh Kumar Chourasia, Vivek Singh, Estimation of photonic band gap in the hollow core cylindrical multilayer structure, 191–199, Copyright 2018, with permission from Elsevier)

Table 4.2 LBE, RBE, and PBG positions shift with the angle of incidence at design wavelengths of 632.8 nm for high R.I. contrast BFs [25]

S.no.	High R.I. contrast BFs at design wavelength 632.8 nm						
1.	θ°	**0**	**15**	**30**	**45**	**60**	**75**
2.	**LBE (nm)**	540.0	537.0	526.0	507.0	489.0	476.0
3.	**RBE (nm)**	761.0	755.0	737.0	712.0	687.0	667.0
4.	**PBG (nm)**	221.0	218.0	211.0	205.0	235.0	210.0

to attain PPBG in the proposed BFs, which yield a high PBG. The core R.I., breadth, and spectral location of the LBE, RBE, and PBG are significantly influenced by the light beam's angle of incidence. Furthermore, dispersion analysis of BFs reveals the optimal structural design opportunity for BFs.

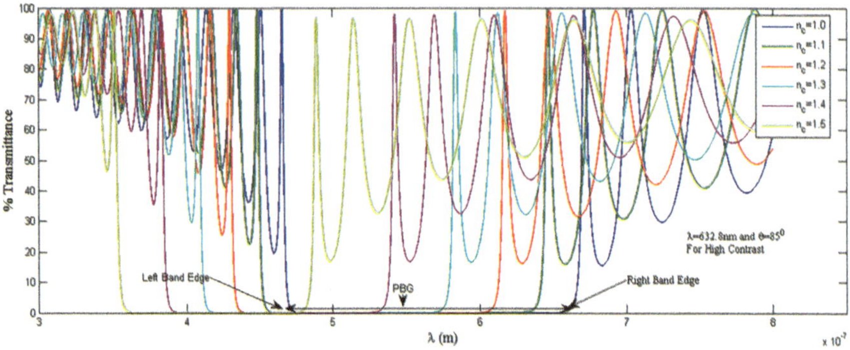

Fig. 4.9 LBE, RBE, and PBG position shift with core refractive indices in high R.I. contrast BFs at design wavelength 632.8 nm. (Reprinted from [25]: Superlattices and Microstructures, 116, Ritesh Kumar Chourasia, Vivek Singh, Estimation of photonic band gap in the hollow core cylindrical multilayer structure, 191–199, Copyright 2018, with permission from Elsevier)

Table 4.3 LBE, RBE, and PBG position shift with core refractive index at design wavelengths of 632.8 nm for high R.I. contrast BFs

S.no.	High R.I. contrast BFs at design wavelength 632.8 nm						
1.	**Core R.I.**	**1.0**	**1.1**	**1.2**	**1.3**	**1.4**	**1.5**
2.	**LBE (nm)**	470.7	454.5	435.7	414.0	388.6	358.3
3.	**RBE (nm)**	660.8	636.4	607.8	574.0	533.0	480.5
4.	**PBG (nm)**	190.1	181.9	172.1	160.0	144.4	122.1

Table 4.4 Cutoff core radius for various possible guided modes in high R.I. contrast BFs [21]

Guided modes @ $\lambda_0 = 632.8$ nm	TE_{01}& TE_{02}	TE_{03}& TE_{04}	TE_{05}& TE_{06}	TE_{07}& TE_{08}	TE_{09}& TE_{010}	TE_{011}& TE_{012}	TE_{013}& TE_{014}	TE_{015}
Cutoff radius (μm) (for high contrast)	0.395	1.005	1.655	2.284	2.934	3.564	4.193	4.823
	0.690	1.340	1.970	2.603	3.249	3.878	4.523	

References

1. P. Yeh, A. Yariv, C.S. Hong, Electromagnetic propagation in periodic stratified media. I. General theory. J. Opt. Soc. Am. **67**, 423–437 (1977)
2. P. Yeh, A. Yariv, Bragg reflection waveguides. Opt. Commun. **19**, 427–430 (1976)
3. P. Yeh, A. Yariv, E. Marom, Theory of Bragg fiber. J. Opt. Soc. Am. **68**, 1196–1201 (1978)
4. N.J. Doran, K.J. Blow, Cylindrical Bragg fibers: A design and feasibility study for optical communications. J. Lightwave Technol. **1**, 588–590 (1983)
5. R.F. Cregan, B.J. Mangan, J.C. Knight, T.A. Birks, P.S.J. Russell, P.J. Roberts, D.C. Allan, Single-mode photonic band gap guidance of light in air. Science **289**, 415–419 (2000)
6. R.F. Cregan, B.J. Mangan, J.C. Knight, T.A. Birks, P.S.J. Russell, P.J. Roberts, Single-mode photonic bandgap guidance of light in air. Science **285**, 1537–1539 (1999)
7. J.L. Auguste, R. Jindal, J.M. Blondy, J. Marcou, B. Dussardier, G. Monnom, D.B. Ostrowsky, B.P. Pal, K. Thyagarajan, 1800 (ps/nm)/km chromatic dispersion at 1.55 μm in dual concentric core fiber. Electron. Lett. **36**, 1689 (2000)

8. A. Argyros, Guided modes and loss in Bragg fibers. Opt. Exp. **10**, 1411–1417 (2002)
9. S. Dasgupta, B.P. Pal, M.R. Shenoy, Design of dispersion compensating Bragg fiber with an ultrahigh figure of merit. Opt. Lett. **30**, 1917–1919 (2005)
10. S. Dasgupta, B.P. Pal, M.R. Shenoy, Nonlinear spectral broadening in solid-core Bragg fibers. J. Lightwave Technol. **25**, 2475–2481 (2007)
11. J. Sakai, Optical power confinement factor in a Bragg fiber: 1. Formulation and general properties. J. Opt. Soc. Am. B **24**, 9–19 (2007)
12. J. Sakai, J. Sasaki, K. Kawai, Optical power confinement factor in a Bragg fiber: 2. Numerical results. J. Opt. Soc. Am. B **24**, 20–27 (2007)
13. J. Sakai, Optical loss estimation in a Bragg fiber. J. Opt. Soc. Am. B **24**, 763–772 (2007)
14. J. Sakai, T. Kimura, Bending loss of propagation modes in arbitrary-index profile optical fibers. Appl. Opt. **17**, 1499–1506 (1978)
15. C.M. de Sterke, I.M. Bassett, Differential losses in Bragg fibers. J. Appl. Phys. **76**, 680–688 (1994)
16. T. Miya, Y. Terunuma, T. Hosaka, T. Miyashita, Ultimate low-loss single-mode fiber at 1.55 μm. Electron. Lett. **15**, 106–108 (1979)
17. S.G. Johnson, M. Ibanescu, M. Skorobogatiy, O. Weisberg, T.D. Engeness, M. Soljacic, S.A. Jacobs, J.D. Joannopoulos, Y. Fink, Low-loss asymptotically single-mode propagation in large-core omni-guide fibers. Opt. Express **9**, 748–779 (2001)
18. N.K. Chourasia, A. Srivastava, V. Kumar, R.K. Chourasia, Optimizing temperature-dependent molar volume fraction of biodiesel fuel in a pseudo-binary mixture through Bragg fiber waveguide sensor having defect layer. Fuel **293**, 120489 (2021)
19. N.K. Chourasia, A. Srivastava, V. Kumar, R.K. Chourasia, Comparative performance study of liquid core cylindrical Bragg fibre waveguide biosensors. Pramana **95**(1), 1–9 (2021)
20. K.J. Rowland, V.S. Afshar, A. Stolyarov, Y. Fink, T.M. Monro, Bragg waveguides with low-index liquid cores, Opt. Express **20**, 48–62 (2011)
21. R.K. Chourasia, V. Singh, *Comparative analysis of the dispersion characteristic of Bragg fibers having different refractive index contrast*, in International Conference on Fibre Optics and Photonics, Th3A. 75 (2016)
22. R.K. Chourasia, S. Prasad, V. Singh, Performance study of a liquid core Bragg fiber sensor in presence of a defect layer. Opto-Electron. Rev. **25**(3), 215–221 (2017)
23. Q. Hang, T. Brastaviceanu, F. Bergeron, J. Olesik, I. Pavlov, T. Ishigure, M. Skorobogatiy, Photonic bandgap Bragg fiber sensors for banding/displacement detection. Appl. Opt. **52**, 6344–6349 (2013)
24. Q. Hang, M. Skorobogatiy, Resonant bio-and chemical sensors using low-refractive-index-contrast liquid-core Bragg fibers. Sensors Actuators B **161**, 261–268 (2012)
25. R.K. Chourasia, V. Singh, Estimation of photonic band gap in the hollow core cylindrical multilayer structure. Superlattice. Microst. **116**, 191–199 (2018)
26. R.K. Chourasia et al., *Optical Properties of Hollow-Core Bragg Fiber Waveguides*, vol 23 (Benthem Science, 2023), pp. 214–236
27. R.K. Chourasia et al., *Photonic Nanostructured Bragg Fuel Adulteration Sensor*, vol 28 (Benthem Science, 2023), pp. 237–264
28. R.K. Chourasia et al., Exploring optical properties in cylindrical polymeric-chalcogenides photonic materials. Mater. Today Proc. **67**(5), 625–631 (2022)
29. L. Shi, W. Zhang, J. Jin, Y.D. Huang, J.D. Peng, *Hollow core Bragg fiber and its application in trace gas sensing*, in Communications and Photonics Conference and Exhibition (ACP), 2010 Asia, IEEE Xplore. https://doi.org/10.1109/ACP.2010.5682586 (2011)
30. M.S. Chen, C.J. Wu, T.J. Yang, Narrowband reflection and transmission filter in an annular defective photonic crystal containing an ultra-thin metallic film. Opt. Commun. **285**, 3143–3149 (2012)
31. N.K. Chourasia, A. Srivastava, V. Kumar, R.K. Chourasia, Doubly electrically tuned cylindrical Bragg fiber waveguide inline optical filter for multiwavelength LASER applications. Mater. Today Commun. **25**, 101620 (2020)

32. R.K. Chourasia, V. Prasad, V. Singh, Voltage-tunable pass band in cylindrical multilayered structure containing PMMA and 0.67 PMN–0.33 PT single crystal as a defect layer. Opt. Quant. Electron. **48**(12), 1–9 (2016)
33. G. Sharma, S. Kumar, S. Prasad, V. Singh, Theoretical modeling of one-dimensional photonic crystal based optical demultiplexer. J. Mod. Opt. (2015). https://doi.org/10.1080/09500340.2015.1088097
34. R.K. Chourasia, V. Singh, *Cylindrical Bragg Waveguide-Based Temperature Sensors*, in IEEE Uttar Pradesh Section International Conference on Electrical, Computer and Electronics Engineering (UPCON), 87-90 (2016)
35. D.R. Smith, R. Dalichaouch, N. Kroll, S. Schultz, S.L. McCall, P.M. Platzman, Photonic band structure without and with defect in one-dimensional photonic crystals. J. Opt. Soc. Am. B. Opt. Phys. **10**, 314–321 (1993)
36. S. Golmohammadi, Y. Rouhani, K. Abbasian, A. Rostami, Photonic bandgaps in quasiperiodic multilayer using Fourier transform of the refractive index profile. Prog. Electromagn. Res. **5**, 133–152 (2008)
37. L. González-García, S. Colodrero, H. Míguez, A.R. González-Elipe, Single-step fabrication process of 1-D photonic crystals coupled to nanocolumnar TiO2 layers to improve DSC efficiency. Opt. Express **23**, A1642–A1650 (2015)
38. Y. Gao, N. Guo, B. Gauvreau, M. Rajabian, O. Skorobogata, E. Pone, O. Zabeida, L. Martinu, C. Dubois, M. Skorobogatiy, Consecutive solvent evaporation and co-rolling techniques for polymer multilayer hallow fiber perform fabrication. J. Mater. Res. **21**, 2246–2254 (2006)
39. E. Pone, C. Dubois, N. Guo, Y. Gao, A. Dupuis, F. Boismenu, S. Lacroix, M. Skorobogatiy, Drawing of the hollow all-polymer Bragg fibers. Opt. Express **14**, 5838–5852 (2006)
40. M. Kolle, A. Lethbridge, M. Kreysing, J.J. Baumberg, J. Aizenberg, P. Vukusic, Bio-inspired band-gap tunable elastic optical multilayer fibers. Adv. Mater. **25**(15), 2239–2245 (2013)
41. A. Abouraddy, M. Bayindir, G. Benoit, et al., Towards multi-material multifunctional fibers that see, hear, sense, and communicate. Nat. Mater. **6**, 336–347 (2007)
42. K. Kuriki, O. Shapira, S.D. Hart, G. Benoit, Y. Kuriki, J.F. Viens, M. Bayindir, J.D. Joannopoulos, Y. Fink, Hollow multilayer photonic bandgap fibers for NIR applications. Opt. Express **12**, 1510–1517 (2004)
43. M. Skorobogatiy, A. Dupuis, Ferroelectric all-polymer hollow Bragg fibers for terahertz guidance. Appl. Phys. Lett. **90**, 113514 (2007)
44. B. Temelkuran, S.D. Hart, G. Benoit, J.D. Joannopoulos, Y. Fink, Wavelength-scalable hollow optical fibres with large photonic band gaps for CO_2 laser transmission. Nature **420**, 650–653 (2002)
45. M.A. Kaliteevski, R.A. Abram, V.V. Nikolaev, G.S. Sokolovski, Bragg reflectors for cylindrical waves. J. Mod. Opt. **46**(5), 875–890 (1999)
46. R.K. Chourasia, C.S. Yadav, A. Upadhyay, N.K. Chourasia, V. Singh, Comparative study of defect mode intensity and wavelength modulation in Bragg fiber waveguide sensors. Optik, 164198 (2020)
47. A.H. Cherin, *An Introduction to Optical Fibers* (Mc Graw-Hill, Tokyo, 1983)
48. M. Abromowitz, I.A. Stegun, *Handbook of Mathematical Functions with Formulas, Graph and Mathematical Tables* (Dover, New York, 1972)
49. M. Born, E. Wolf, *Principles of Optics* (Cambridge University Press, Cambridge/London, 1999)

Chapter 5
Optical Properties of Asymmetrical Bragg Fiber: With Defect

Abstract In the previous chapter, it was observed that the hollow-core high-refractive index contrast of cladding BFs shows high order PPBG along with FWHM at the considered visible design wavelength window 632.8 nm supporting communication as well sensing technology. We have also observed a considerable shift in LBE, RBE, and the whole PPBG with the shift in incidence angle, no. of bilayer periodicity as well as core refractive index. Hence, it also opens the necessity of studying low refractive index contrast cladding BFs for lowering significantly the width of PPBG for considering the obtained lobes as sensing signals. It also suggests on behalf of lowering FWHM of the signal that we can also opt for modification in the present BFs structure. The present chapter will discuss the comparative study between low and high refractive index contrast BFs and explore the interesting result obtained. Along with this, introducing the defect layer either by breaking the symmetry of periodic bilayers in BFs or adding an additional layer in between the periodic cladding layers. So, the present chapter deals with the design mechanism of hollow-core BFs with defects and matching the electric and magnetic fields at various interfaces and cylindrical slabs, the optical relation has been stabilized to obtain interesting features of such multilayer asymmetrical structure. The defect layer within the cylindrical periodic structure exhibits a peak in the ideal photonic band gap region. The width of the defect peak at half of its maximum intensity is determined by the periodicity of the cladding layers. Additionally, the spectral location and displacement of the defect's peak are contingent upon the light's angle of incidence, the refractive index of the core material, and the structure's designated wavelength. Thus, it is more suitable to regard the peak due to the defect to be a sensing signal rather than taking the entire photonic band gap as a signal for sensing.

5.1 Introduction

COFs enable the propagation of electromagnetic waves (EMWs) by total internal reflection (TIR). In order to accomplish total internal reflection (TIR) in COFs, it is necessary for the refractive index (RI) of the core to be greater than the refractive

R. K. Chourasia, A. Katti, *Bragg Fibers*,
https://doi.org/10.1007/978-3-031-65164-9_5

index of the cladding. The high refractive index (RI) of the core material leads to the occurrence of optical losses, dispersion, nonlinear effects, birefringence, and other unwanted phenomena. By directing the light via hollow-core fibers, these effects are reduced or eliminated. Using the Bragg reflections process, multilayer concentric cylindrical layers may be used to build hollow-core fibers. The claddings of these fibers are composed of photonic crystals in the shape of cylindrical dielectric mirrors. You may calculate the reflectance via a multilayer structure by using Fresnel's reflection, which is caused by constructive interference of the reflected wave by means of these alternating dielectric mirrors. Such multilayer reflections are described by Bragg's law and are comparable to X-ray scatterings in crystal structure. Thus, Bragg fibers (BFs) are fibers that have periodic multilayer claddings with alternating permittivity over a core material that is either hollow or has a low refractive index. Researchers first began looking at the structure of BFs in 1970, and the notion of BFs was first presented in 1976 [1]. This research uncovered a fascinating property of such a structure: it admits just a specific range of wavelengths into its interior. Given their unique characteristics and hollow-core design, these fiber architectures show promise for high-power laser transmission applications where nonlinear effects are largely mitigated [2]. Other potential uses for similar devices include bio, chemical, fuel adulteration sensors [3–12], strain [13], optical de-multiplexers [14–16], omni reflectors, telecommunication systems [17–21], and narrow band transmission filters [14, 15]. The mirror characteristics of these claddings are defined by the constructive interference of reflected waves and Fresnel reflection from several surfaces of the multilayer structure with varying permittivity values. Bragg's condition describes these reflections from multilayer dielectric mirrors, which are similar to X-ray scatterings in the crystal structure. Therefore, "Bragg fiber" is the name given to fibers that have multilayer periodic claddings over a core material that is optically less dense. In 1976, a team of academics put forth the initial idea of a Bragg fiber, and subsequent studies on the topic started in the 1970s. The unique quality of these fibers is that they permit the propagation of a limited range of wavelengths within the structure's interior. Fibers with these unique characteristics have recently attracted interest as a possible medium for high-power laser transmission with greatly reduced nonlinear effects. Additionally, they show promise in a number of fields and devices, including bio, chemical, and strain sensing; communications; optical de-multiplexers; and narrow band transmission filters [3–11]. Nonetheless, scientists have been interested in creating, studying, and improving the optical characteristics of one-dimensional planar photonic crystals, which also serve as the fundamental periodic arrangement of materials with high and low refractive indices [11, 12]. 1-D photonic crystals [22] have recently attracted a lot of attention from academics because of how simple they are to make—just one step [23]. A defect cavity has been created in the structure of one-dimensional planer photonic crystals [22] by researchers who have broken the symmetry of the multilayer structure. The photonic band gap (PBG) area has a passband with a reduced full width at half maximum (FWHM) because of the Febry-Perot resonator cavity that has been incorporated into the multilayer planar constructions. The researchers also discovered that inline optical applications like narrowband transmission filters and high-power laser systems are not possible with planar photonic crystal structures. For this reason, multilayer cylindrical systems with concentric cores and low refractive indices should also be considered in the current investigation. We have selected a hollow-core BFs construction with an air defect cavity for our investigation. Theoretical research and design towards a PBG area transmission window of 100% is of interest to us. Using drawing and rolling procedures, researchers have been involved in constructing such structures recently [15, 16]. Within the photonic band gap (PBG) area, this cavity functions as a Fabry-Pérot resonant cavity, producing a passband, also known as a defect peak. The inline optical applications of these planar photonic crystals, such as narrow-band transmission filters and high-power laser systems, are not feasible. In order to achieve a 100% transmission window in the PBG region, the current this chapter studies and designs BFs—concentric multilayer cylindrical systems—with a faulty air cavity. As a result, BFs with high or low refractive index contrasts are easily fabricatable. The fabrication of such fiber using drawing and rolling processes has recently captivated researchers [24–27]. They used fiber drawing towers to draw the bilayer preform to length in this method. Using this technique, high-contrast BFs with a polymer low index layer and a chalcogenide or perovskite high index layer may be created [3]. Nevertheless, low-contrast BFs may be made using the rolling process; materials for both the high and low index can be polymer films, such as polystyrene (PS) and polymethyl methacrylate (PMMA) [4]. Any number of optoelectronic uses can make use of either kind of fiber resonator. The current researchers are curious about the optical characteristics of high- and low-contrast BFs with a flaw cavity because of their rising popularity. It is reasonable to examine the behavior of the defect peak in both low- and high-contrast BFs, since in these structures the EM wave guiding happens through the Bragg reflection mechanism and the peak develops because of a defect cavity. To investigate the propagation properties of the suggested BFs, we use the reliable transfer matrix approach.

5.2 Introducing Defects in a Periodic Structure: Proposed Model

Figure 5.1 illustrates a cross-sectional image of the hollow-core BFs with a defect cavity, showing the internal structure along its length. The figure displays the refractive index (RI) and radius of the core, denoted respectively as n_c, r_0. Additionally, n_H, n_L and d_H, d_L represent the R.I.s and thicknesses of the high and low-index photonic layers, respectively. These layers combine to form a bilayer structure. A multilayer of bilayers has been constructed by repeating these layers of BFs. Furthermore, a defect cavity with a refractive index of n_D and a thickness of d_D is inserted into the center of the multilayer Bragg filters, serving as a Fabry-Pérot resonant cavity. Within BFs, the EM wave is expected to possess a cylindrical shape as a result of the cylindrical symmetry inherent in the proposed structure. The

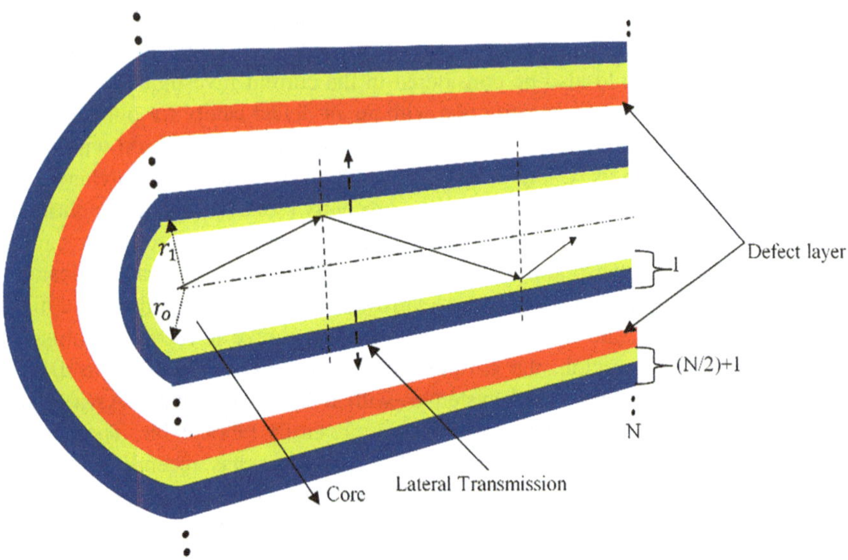

Fig. 5.1 BFs structure with defect cavity cross-sectional view along the length

cylindrical wave is presumed to be radiating away from the axis of symmetry $r = 0$, and then impinges against the initial cylindrical contact $r = r_0$. We can utilize the Transfer Matrix Method (TMM) for the cylindrical waves [28] to compute the reflectance of the entire structure.

All fields are considered to have a temporal component of $\exp(i\omega t)$. If the medium is considered devoid of sources, Maxwell's equations [29], can be expressed as follows:

$$\nabla \times E = -j\omega H \tag{5.1a}$$

$$\nabla \times H = j\omega E \tag{5.1b}$$

The nonzero fields E_z, H_θ, H_r in cylindrical coordinates (r, θ, z) [29] that represent TE mode or H polarization satisfy the subsequent three equations in each layer:

$$\frac{1}{r}\frac{\partial E_z}{\partial \theta} = -j\omega\mu H_r \tag{5.2a}$$

$$\frac{\partial E_z}{\partial r} = j\omega\mu H_\theta \tag{5.2b}$$

$$\frac{\partial (rH_\theta)}{\partial r} - \frac{\partial H_r}{\partial \theta} = -j\omega\varepsilon r E_z. \tag{5.2c}$$

Equation (5.2c) can be used to write the differential equation governing the tangential component of the electric field E_z as follows:

$$r\frac{\partial}{\partial r}\left(r\frac{\partial E_z}{\partial r}\right)-r^2\frac{1}{\mu}\frac{\partial \mu}{\partial r}\frac{\partial E_z}{\partial r}+\frac{\partial}{\partial \theta}\left(\frac{\partial E_z}{\partial \theta}\right)+\omega^2\varepsilon\mu r^2 E_z=0. \tag{5.3}$$

Let's now assume that the solution U and the field component $E_z = U(r)\Theta(\theta) = U(r)e^{im\theta}$, are provided as,

$$U(r)=C_1J_m(kr)+C_2Y_m(kr), \tag{5.4}$$

where C_1 and C_2 are constants, J_m is a Bessel function, Y_m is the Neumann function, and k is the wave number in the material medium. The azimuthal mode number, denoted by the subscript "m", is zero in this investigation. Similarly, the magnetic field's tangential component can be represented as, $H_\theta = V(r)e^{im\theta}$, where $V(r)$ is given by,

$$V(r)=-jp\left(C_1J'_m(kr)+C_2Y'_m(kr)\right), \tag{5.5}$$

where the material medium's intrinsic admittance is denoted by $p=\sqrt{\varepsilon/\mu}$.

A matrix for only one layer is created by concurrently applying Eqs. (5.4 and 5.5) to relate the electric and magnetic fields at the two interfaces. For example, the matrix $\widehat{T}_1$ representing the first layer at the interface ($r=r_0$ and $r=r_1$) is denoted as [28, 30]:

$$\begin{bmatrix} U(r_1) \\ V(r_1) \end{bmatrix}=\widehat{T}_1\begin{bmatrix} U(r_0) \\ V(r_0) \end{bmatrix} \tag{5.6}$$

The mono-layer matrix $\widehat{T}_1=\begin{bmatrix} t_{11} & t_{12} \\ t_{21} & t_{22} \end{bmatrix}$ components can be written as,

$$t_{11}=\frac{\pi}{2}k_1r_0\left[Y'_m(k_1r_0)J_m(k_1r_1)-J'_m(k_1r_0)Y_m(k_1r_1)\right], \tag{5.7a}$$

$$t_{12}=j\frac{\pi}{2}\frac{k_1}{p_1}r_0\left[J_m(k_1r_0)Y_m(k_1r_1)-Y_m(k_1r_0)J_m(k_1r_1)\right], \tag{5.7b}$$

$$t_{21}=-j\frac{\pi}{2}k_1r_0p_1\left[Y'_m(k_1r_0)J'_m(k_1r_1)-J'_m(k_1r_0)Y'_m(k_1r_1)\right], \tag{5.7c}$$

$$t_{22}=\frac{\pi}{2}k_1r_0\left[J_m(k_1r_0)Y'_m(k_1r_1)-Y_m(k_1r_0)J'_m(k_1r_1)\right], \tag{5.7d}$$

where $p_1=\sqrt{\varepsilon_1/\mu_1}$. Equation (5.7) demonstrates that the matrix elements are directly influenced by the two interfaces' radii. Due to the introduction of an extra defect layer in the BFs using Eq. (5.7), the TMM for the defect layer will be produced by substituting $r_0\rightarrow r_{D1}$, $r_1\rightarrow r_{D2}$, $k_1\rightarrow k_D$ and $p_1\rightarrow p_D=\sqrt{\varepsilon_D/\mu_D}$. Here r_{D1} and r_{D2} are the position of the interfaces of the defect cavity, and k_d and p_d are wave vectors associated with the defect cavity and characteristic impedance of the defect cavity, respectively. The component of the defect cavity matrix can be written as $\widehat{T}_{\mathrm{D}} = \begin{bmatrix} t_{\mathrm{D}11} & t_{\mathrm{D}12} \\ t_{\mathrm{D}21} & t_{\mathrm{D}22} \end{bmatrix}$

$$t_{\mathrm{D}11} = \frac{\pi}{2} k_{\mathrm{D}} r_{\mathrm{D}1} \left[Y'_m(k_{\mathrm{D}} r_{\mathrm{D}1}) J_m(k_{\mathrm{D}} r_{\mathrm{D}2}) - J'_m(k_{\mathrm{D}} r_{\mathrm{D}1}) Y_m(k_{\mathrm{D}} r_{\mathrm{D}2}) \right], \tag{5.8a}$$

$$t_{\mathrm{D}12} = j\frac{\pi}{2} \frac{k_{\mathrm{D}}}{p_{\mathrm{D}}} r_{\mathrm{D}1} \left[J_m(k_{\mathrm{D}} r_{\mathrm{D}1}) Y_m(k_{\mathrm{D}} r_{\mathrm{D}2}) - Y_m(k_{\mathrm{D}} r_{\mathrm{D}1}) J_m(k_{\mathrm{D}} r_{\mathrm{D}2}) \right], \tag{5.8b}$$

$$t_{\mathrm{D}21} = -j\frac{\pi}{2} k_{\mathrm{D}} r_{\mathrm{D}1} p_{\mathrm{D}} \left[Y'_m(k_{\mathrm{D}} r_{\mathrm{D}1}) J'_m(k_{\mathrm{D}} r_{\mathrm{D}2}) - J'_m(k_{\mathrm{D}} r_{\mathrm{D}1}) Y'_m(k_{\mathrm{D}} r_{\mathrm{D}2}) \right], \tag{5.8c}$$

$$t_{\mathrm{D}22} = \frac{\pi}{2} k_{\mathrm{D}} r_{\mathrm{D}1} \left[J_m(k_{\mathrm{D}} r_{\mathrm{D}1}) Y'_m(k_{\mathrm{D}} r_{\mathrm{D}2}) - Y_m(k_{\mathrm{D}} r_{\mathrm{D}1}) J'_m(k_{\mathrm{D}} r_{\mathrm{D}2}) \right]. \tag{5.8d}$$

Therefore, it is possible to easily create the matrix for every existing layer of stacked BFs. To generate the total transfer matrix $\widehat{T}$ that relates the initial and final interfaces in the BFs structure, a total of 2 N matrices, together with the defect cavity matrix, are required,

$$\begin{bmatrix} U(r_{\mathrm{f}}) \\ V(r_{\mathrm{f}}) \end{bmatrix} = \widehat{T} \begin{bmatrix} U(r_0) \\ V(r_0) \end{bmatrix}, \tag{5.9}$$

where

$$\widehat{T} = \begin{bmatrix} T_{11} & T_{12} \\ T_{21} & T_{22} \end{bmatrix} = \widehat{T}_{2\mathrm{N}} \widehat{T}_{2\mathrm{N}-1} \ldots\ldots \widehat{T}_{\mathrm{N}+1} \widehat{T}_{\mathrm{D}} \widehat{T}_{\mathrm{N}} \ldots\ldots \widehat{T}_2 \widehat{T}_1. \tag{5.10}$$

By utilizing this computed transfer matrix, one can ascertain the reflection and transmission coefficients using the subsequent relationships [28],

$$r_d = \frac{\left(T'_{21} + jp_0 C^{(2)}_{m0} T'_{11}\right) - jp_f C^{(2)}_{mf} \left(T'_{22} + jp_0 C^{(2)}_{m0} T'_{12}\right)}{\left(-jp_0 C^{(1)}_{m0} T'_{11} - T'_{21}\right) - jp_f C^{(2)}_{mf} \left(-jp_0 C^{(1)}_{m0} T'_{12} - T'_{22}\right)}, \tag{5.11}$$

$$t_d = \frac{4\sqrt{\varepsilon_0/\mu_0}}{\pi K r_0 H^{(2)}_m(k_0 r_0) H^{(1)}_m(k_0 r_0) \left[\left(-jp_0 C^{(1)}_{m0} T'_{11} - T'_{21}\right) - jp_f C^{(2)}_{mf} \left(-jp_0 C^{(1)}_{m0} T'_{12} - T'_{22}\right) \right]}, \tag{5.12}$$

where $p_0 = \sqrt{\varepsilon_0/\mu_0}$ and $p_{\mathrm{f}} = \sqrt{\varepsilon_{\mathrm{f}}/\mu_{\mathrm{f}}}$ are the admittance of core and final media, $T'_{11}, T'_{12}, T'_{21}$ and T'_{22} are the matrix element of the inverse matrix of $\hat{T}$, $K = \omega\sqrt{\varepsilon_0 \mu_0}$ is the free-space wave number and

$$C_{ml}^{(1,2)} = \frac{H_m^{(1,2)\prime}(k_l r_l)}{H_m^{(1,2)}(k_l r_l)}, l = 0, f. \tag{5.13}$$

where $H_m^{(1)}$ and $H_m^{(2)}$ are the Hankel functions of the first and second kind. Using Eqs. (5.11) and (5.12), the percentage reflectance %R and Transmittance %T are:

$$\%R = 100 \times |r_d|^2, \%T = 100 \times \frac{n_{\text{out}}}{n_{\text{c}}} |t_d|^2. \tag{5.14}$$

where the refractive indexes of the core and outer media are respectively n_{c} and n_{out}.

An evaluation of the performance of the proposed BFs structures is necessary, which requires the analysis of multiple sensing metrics. Key factors include sensitivity, detection accuracy, and quality metrics. The sensitivity of the sensor is determined by the quotient of the output measured by the sensor and the change in the physical quantity being measured. The sensitivity of the sensor is mathematically described as the ratio of the spectral shift in resonance wavelength to the very small change in refractive index (R.I.) of the analyte present in the core.

$$S = \Delta\lambda_{\text{res}} / \Delta n_{\text{c}}. \tag{5.15}$$

Detection accuracy refers to the degree of precision in determining the true value of the measurand. The detection accuracy of this sensor is determined by the ratio of the change in resonance wavelength to the full width at half maximum (FWHM) of the defect peak. The quality parameter provides a comprehensive assessment of the sensor's performance in terms of its sensitivity and accuracy in detecting signals.

5.3 Obtained Outcomes

The study focuses on the propagation characteristics of hollow-core high and low refractive index contrast cladding Bragg fibers with defect cavities in the visible region. The spectral position of the defect peak is determined by the wavelength at which the operation occurs. The design wavelength, denoted as A, is well-suited for both communication and sensing purposes. The fiber core and outer layer must be considered as air, denoted as $n_{\text{c}} = n_{\text{out}} = 1.0$, and the core dimension must be sufficiently small, denoted as $r_0 = 100(d_H + d_L)$, in order to reduce losses caused by higher-order modes. The cladding of the hollow-core high-contrast BFs consists of bilayers made of high refractive index (RI) chalcogenide material (As_2Se_3) with a refractive index of $n_H = 2.82$, and low RI polymer material (PEI) with a refractive index of $n_L = 1.66$. These bilayers are arranged in a multilayer structure, with an air defect cavity located in the middle of the claddings [3, 31]. The thicknesses of the bilayers are determined by the quarter wave stack condition, with the high RI bilayers having a thickness of $d_H = \lambda_d/4n_H$ and the low RI bilayers having a

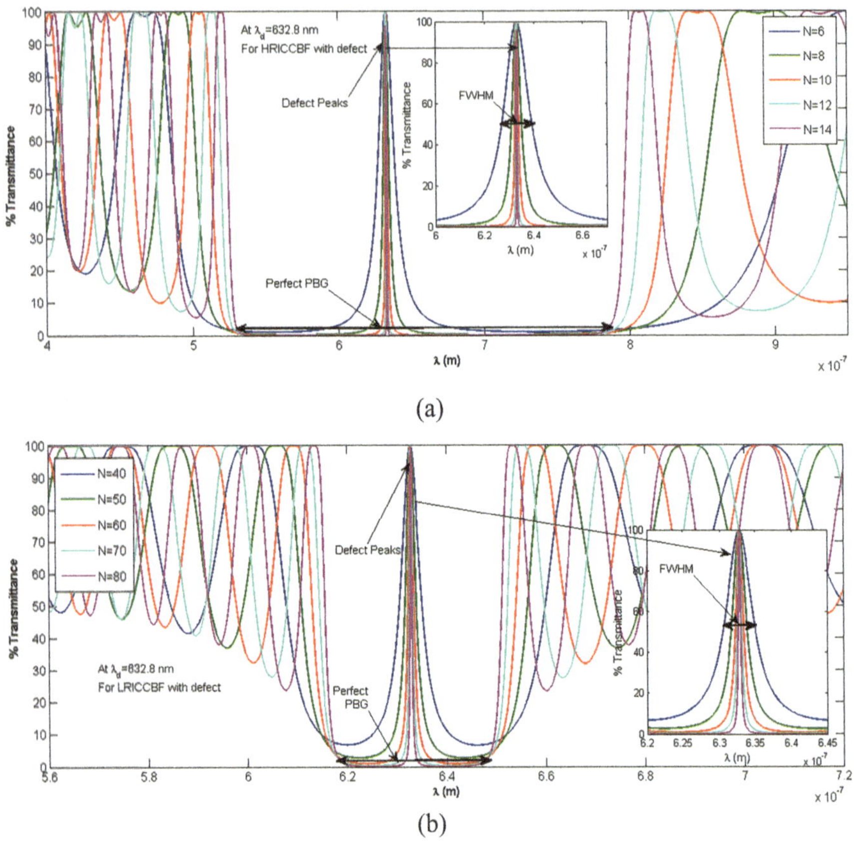

Fig. 5.2 Effect of the number of periodic claddings on FWHM of the defect signals and PBG at the design wavelength 632.8 nm for hollow-core (**a**) high-contrast BFs (**b**) Low-contrast BFs with defect cavity. (Reprinted from Optik, 200, Ritesh Kumar Chourasia, Chandan Singh Yadav, Abhishek Upadhyay, Nitesh Kumar Chourasia, Vivek Singh, Analysis of Bragg fiber waveguides having a defect layer for biosensing application, 163,400, Copyright 2020, with permission from Elsevier)

thickness of $d_L = \lambda_d/4n_L$, where λ is the wavelength and d is the desired thickness of the claddings. The bilayers cladding of hollow-core low refractive index (RI) Bragg fibers (BFs) consist of a high refractive index polymer material (PS) with a refractive index of 1.581 and a low refractive index polymer (PMMA) with a refractive index of 1.487. The claddings also contain an air defect cavity positioned in the middle of the multilayer structure [4]. The thicknesses of the claddings follow the same quarter-wave stack condition as mentioned earlier.

Figure 5.2 demonstrates the presence of PPBG in both of the examined BFs structures at the designated design wavelength. The PPBG refers to the wavelength range in which there is no transmittance, meaning that no light is able to pass through. The figures, labeled 5.2a, b, display the transmittance of hollow-core high

Table 5.1 Variation in FWHM and PPBG with number of unit cells (N) in hollow-core high- and low R.I.contrast BFs with defect cavity

	High-contrast BFs with defect cavity		Low-contrast BFs with defect cavity	
	At design wavelength 632.8 nm			
S. no.	N	FWHM (nm)	N	FWHM (nm)
1.	6	12.2	40	4.6
2.	8	3.9	50	2.3
3.	10	1.4	60	1.2
4.	12	0.5	70	0.6
5.	14	0.2	80	0.4
PPBG	259 nm (at N = 14)		31 m (at N = 80)	

and low refractive index contrast Bragg fibers with an air defect cavity at the design wavelength of 632.8 nm. The measured peak-to-peak bandgap for high refractive index contrast Bragg filters is 259 nm, while for low refractive index contrast Bragg filters it is 31 nm. The measured PPBG in high-contrast BFs is consistently greater than that in low-contrast BFs at the designated wavelength, owing to the device structure's pronounced reflectivity. Both figures clearly demonstrate that a small number of periodic cladding layers (N = 14) are enough to achieve the photonic bandgap (PPBG) in hollow-core high-contrast Bragg fibers with an air defect cavity. On the other hand, a large number of periodic cladding layers (N = 80) are necessary for obtaining the PPBG in hollow-core low-contrast Bragg fibers. Furthermore, the inclusion of an air defect cavity in these Bragg fibers results in a highly distinct pass band (defect signal) in the PPBG region, exhibiting nearly 100% transmittance as a consequence of the presence of the Fabry-Pérot resonant condition. The full width at half maximum (FWHM) of the narrow frequency range narrows progressively as the cladding period N increases, as shown in Table 5.1. A smaller FWHM value improves the sensing performance of the BFs sensor and optimizes the quality parameter of the inline Bragg filter beam. Based on the preceding analysis, it is evident that the photonic band gap (PPBG) of the suggested Bragg fiber structures is greatly influenced by the refractive index (RI) contrast of the materials employed. As the RI contrast decreases, the PPBG also decreases. Additionally, the position and size of the band gap are not affected by multiple layers, but the full width at half maximum (FWHM) of the defect peak is directly dependent on the number of periodic bilayers.

In order to observe the impact of different angles at which EM waves intersect with photonic bandgap structures (PPBG), a total of 14 unit cells are utilized in hollow-core high-contrast BFs with a defect cavity. Similarly, a total of 80 unit cells are employed in hollow-core low-contrast BFs with a defect cavity. Figure 5.3a, b display the spectrum of PPBG and the change in defect signal at various angles of incidence. These figures illustrate the specific location of the defect signal and the PPBG for both hollow-core high-contrast and low-contrast BFs, considering a

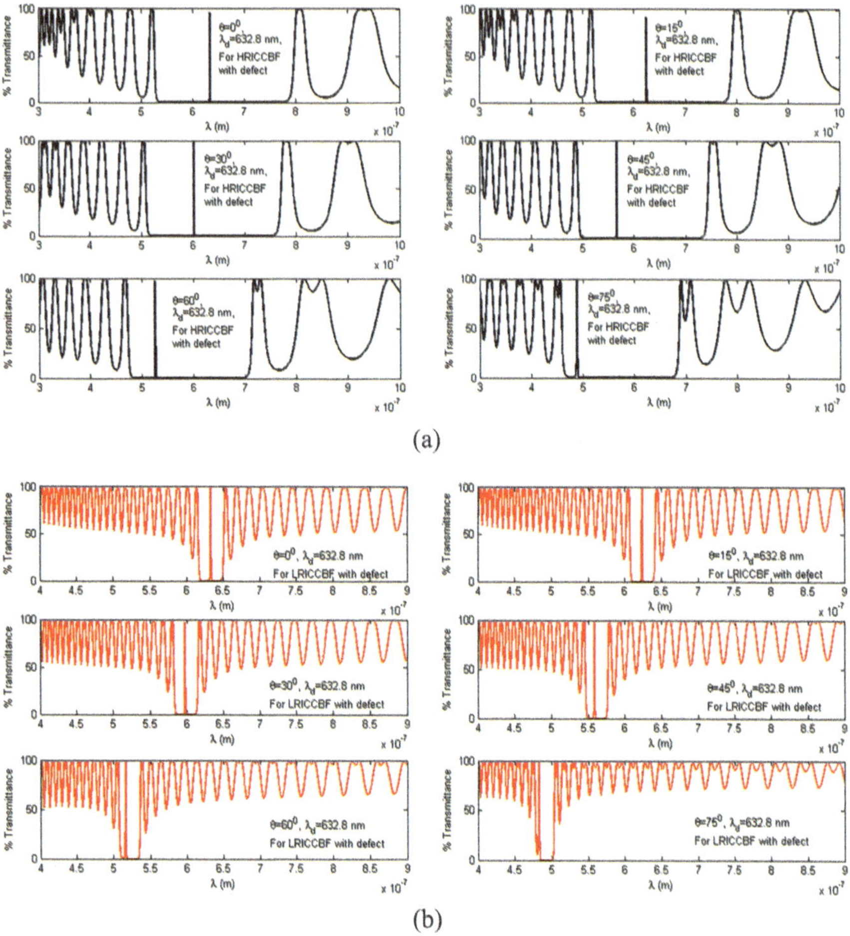

Fig. 5.3 Spectral shift of defect signal and obtained PPBG with angle of incidence in (**a**) high-contrast BFs and (**b**) low-contrast BFs at design wavelength 632.8 nm. (Reprinted from Optik, 200, Ritesh Kumar Chourasia, Chandan Singh Yadav, Abhishek Upadhyay, Nitesh Kumar Chourasia, Vivek Singh, Analysis of Bragg fiber waveguides having a defect layer for biosensing application, 163400, Copyright 2020, with permission from Elsevier)

wavelength of 632.8 nm. The defect signal and PPBG positions are determined from these figures and recorded in Table 5.2. Table 5.2 demonstrates that a change in angle of incidence from 0° to 75° can cause a shift in the defect peak of 145 nm and the photonic bandgap (PPBG) of 43.2 nm for a design wavelength of 632.8 nm in hollow-core high-contrast Bragg fibers (BFs). In hollow-core low-contrast BFs, a change in angle of incidence results in a shift of the defect signal position by 150.1 nm and a change in the photonic bandgap by 6.8 nm at a design wavelength of 632.8 nm. The observed reductions in PBG with incidence angles are attributed to

Table 5.2 Variation in the defect signal positions and PPBG with angle of incidence

	High-contrast BFs with defect cavity			Low-contrast BFs with defect cavity	
	At design wavelength 632.8 nm				
S. no.	$\theta°$	DPP (nm)	PPBG (nm)	DPP (nm)	PPBG (nm)
1.	0	632.8	258.6	632.8	31.1
2.	15	624.6	256.6	623.3	30.7
3.	30	600.8	250.2	596.7	29.3
4.	45	565.6	240.5	558.1	27.3
5.	60	525.0	228.9	516.3	25.0
6.	75	487.8	215.4	482.7	24.3
Shift at $\theta = 75°$ concerning $\theta = 0°$		145.0	43.2	150.1	6.8

a decrease in the geometric path of light. Moreover, the defect signals exhibit a blue shift when the geometrical route of light decreases.

The hollow cores of these structures include a BF arrangement with an air defect cavity. This configuration can be utilized for chemical and bio-sensing purposes by considering the defect signal within the photonic bandgap (PPBG) as a sensing signal. Figure 5.4 illustrates the change in defect signal position with core refractive indices for both high- and low-contrast BFs at the specified design wavelengths. The corresponding data is also included in Table 5.3. Table 5.3 demonstrates that in sensing applications, one can observe and track the change in spectral frequency of the defect signal found in the photonic bandgap structure. Figure 5.2 clearly demonstrates that the Full Width at Half Maximum (FWHM) of the defect peak is significantly less than the FWHM of the Photonic Bandgap (PPBG). Consequently, when utilizing the defect signal as a sensing signal, these parameters exhibit substantial values. Therefore, it has been determined that monitoring defect peaks, rather than monitoring PPBG, will yield superior detection accuracy and quality parameters for the sensor systems. In low-contrast BFs structure, the FWHM of the defect signal is smaller compared to the FWHM of high-contrast BFs structure. Thus, it is advisable to utilize a low-contrast Bragg-Fresnel structure with a defect cavity for sensing purposes.

5.4 Discussions and Conclusion

Extensive analysis has been conducted on periodic claddings with high and low contrast, as well as hollow-core Bragg fibers that contain a defect cavity. These structures were thoroughly investigated and subsequently compared at the specific wavelength of 632.8 nm. Comparative investigations demonstrate that a smaller number of periodic bilayers are needed to achieve high-contrast BFs with PPBG

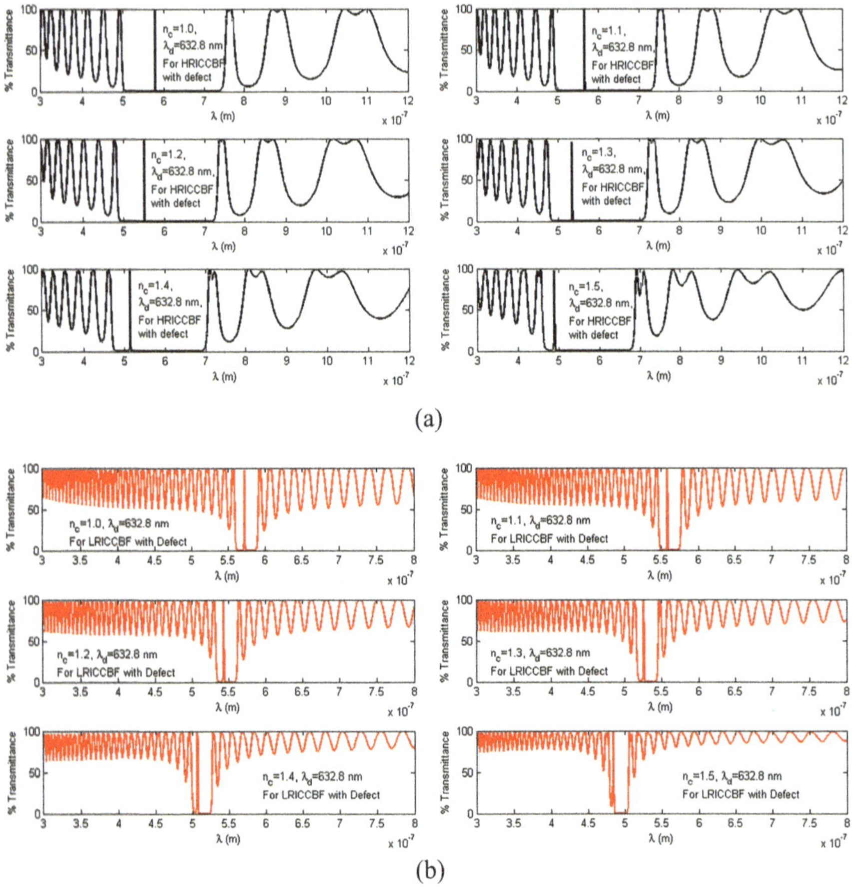

Fig. 5.4 Variation in photonic band with core R.I.s in (**a**) high-contrast BFs (**b**) low-contrast BFs at design wavelength 632.8 nm. (Reprinted from Optik, 200, Ritesh Kumar Chourasia, Chandan Singh Yadav, Abhishek Upadhyay, Nitesh Kumar Chourasia, Vivek Singh, Analysis of Bragg fiber waveguides having a defect layer for biosensing application, 163400, Copyright 2020, with permission from Elsevier)

compared to low-contrast BFs. A distinct defect peak is found in the reported PPBG of both high- and low-contrast Bragg fibers due to the presence of a defect cavity that functions as a Fabry-Pérot resonator. In addition, it has been noted that the full width at half maximum (FWHM) of the defect peak reduces as the periodicity of the cladding layers increases while keeping the design wavelength constant. The change in the positions of defect peaks and photonic bandgap has also been noticed with the angle at which light strikes the surface and the alteration of the refractive index of the core for both types of Bragg fiber structures at the specified design wavelength. The results confirm that when the angle of incidence varies, there is a more significant change in the defect and PPBG position for low-contrast BFs compared to high-

Table 5.3 Variation in the defect signal positions and PPBG with core R.I

	High-contrast BFs with defect cavity			Low-contrast BFs with defect cavity	
	At design wavelength 632.8 nm				
S. no.	n_c	DPP (nm)	PPBG (nm)	DPP (nm)	PPBG (nm)
1.	1.0	578.3	244.1	571.9	28.2
2.	1.1	565.6	240.5	558.1	27.4
3.	1.2	551.0	236.3	542.7	26.6
4.	1.3	534.1	231.3	525.3	25.4
5.	1.4	514.0	225.0	505.6	24.7
6.	1.5	488.7	215.7	483.3	23.6
Shift at core RI = 1.5		89.6	28.4	88.6	4.6

Reprinted from Optik, 200, Ritesh Kumar Chourasia, Chandan Singh Yadav, Abhishek Upadhyay, Nitesh Kumar Chourasia, Vivek Singh, Analysis of Bragg fiber waveguides having a defect layer for biosensing application, 163,400, Copyright 2020, with permission from Elsevier

contrast BFs. However, due to the difference in core refractive index, the observed displacement of defect position as well as photonic bandgap is more pronounced in high-contrast Bragg fibers compared to low-contrast Bragg fibers. Thus, we can deduce that the sensitivity of high-contrast BFs is consistently greater than that of low-contrast BFs. However, it is important to note that the full width at half maximum (FWHM) of the defect peak in low contrast BFs is significantly smaller than that of the corresponding high-contrast BFs. Consequently, there was a significant enhancement in the overall excellence of low-contrast BFs. Therefore, low-contrast Bragg filters with defect cavities are strongly recommended for bio-sensing applications.

References

1. P. Yeh, A. Yariv, Bragg reflection waveguides. Opt. Commun. **19**, 427–430 (1976)
2. B. Temelkuran, S.D. Hart, G. Benoit, J.D. Joannopoulos, Y. Fink, Wavelength scalable hollow optical fibers with large photonic bandgaps for CO_2 laser transmission. Nature **420**, 650–653 (2002)
3. K.J. Rowland, V.S. Afshar, A. Stolyarov, Y. Fink, T.M. Monro, Bragg waveguides with low index liquid cores. Opt. Express **20**, 48–62 (2011)
4. R.K. Chourasia, C.S. Yadav, A. Upadhyay, N.K. Chourasia, V. Singh, Comparative study of defect mode intensity and wavelength modulation in Bragg fiber waveguide sensors. Optik **204**, 164198 (2020)
5. R.K. Chourasia, C.S. Yadav, A. Upadhyay, N.K. Chourasia, V. Singh, Analysis of Bragg fiber waveguides having a defect layer for biosensing application. Optik **200**, 163400 (2020)
6. N.K. Chourasia, A. Srivastava, V. Kumar, R.K. Chourasia, Comparative performance study of liquid core cylindrical Bragg fibre waveguide biosensors. Pramana J. Phys. **95**, 9 (2021)
7. Q. Hang, M. Skorobogatiy, Resonant bio-and chemical sensors using low-refractive-index contrast liquid-core Bragg fibers. Sensors Actuators B **161**, 261–268 (2012)

8. D. Passaro, M. Foroni, F. Poli, A. Cucinotta, S. Selleri, J. Laegsgaard, A.O. Bjarklev, All silica hollow core micro structured Bragg fibers for biosensor application. IEEE J. Sensors **8**, 1280–1286 (2008)
9. N.K. Chourasia, A. Srivastava, V. Kumar, R.K. Chourasia, Optimizing temperature-dependent molar volume fraction of biodiesel fuel in a pseudo-binary mixture through Bragg fiber waveguide sensor having defect layer. Fuel **293**, 120489 (2021)
10. R.K. Chourasia, N.K. Chourasia, A. Srivastava, N. Bihari, *Photonic Nanostructured Bragg Fuel Adulteration Sensor*, vol 28 (Benthem Science, 2023), pp. 237–264
11. N.K. Chourasia, N. Bihari, R.K. Chourasia, An optofluidic Bragg fiber sensor for estimating adulterants in a temperature-dependent molar fraction of hydrated mono-alcohol fuels. Heliyon **8**(9), e10532 (2022)
12. L. Shi, W. Zhang, J. Jin, Y.D. Huang, J.D. Peng, *Hollow core Bragg fiber and its application in trace gas sensing*, in Communications and Photonics Conference and Exhibition (ACP) (2010) Asia, IEEE Xplore. https://doi.org/10.1109/ACP.2010.5682586, (2011)
13. Q. Hang, T. Brastaviceanu, F. Bergeron, J. Olesik, I. Pavlov, T. Ishigure, M. Skorobogatiy, Photonic bandgap Bragg fiber sensors for banding/displacement detection. Appl. Opt. **52**, 6344–6349 (2013)
14. G. Sharma, S. Kumar, S. Prasad, V. Singh, Theoretical modeling of one-dimensional photonic crystal based optical de-multiplexer. J. Mod. Opt. (2015). https://doi.org/10.1080/09500340.2015.1088097
15. N.K. Chourasia, A. Srivastava, V. Kumar, R.K. Chourasia, Doubly electrically tuned cylindrical Bragg fiber waveguide inline optical filter for multiwavelength laser applications. Mater. Today Commun. **25**, 101620 (2020)
16. R.K. Chourasia, S. Prasad, V. Singh, Voltage-tunable pass band in a cylindrical multilayered structure containing PMMA and 0.67PMN–0.33PT single crystal as a defect layer. Opt. Quant. Electron. **48**, 539 (2016)
17. Y. Fink, J.N. Winn, S. Fan, C. Chen, J. Michel, J.D. Joannopoulos, E.L. Thomas, A dielectric omnidirectional reflector. Science **282**, 1679–1682 (1998)
18. N. Bihari, A. Srivastava, N.K. Chourasia, R.K. Chourasia, H-polarised EM-wave transport in polymeric-chalcogenide columnar photonic materials. J. Phys. Conf. Ser. **2426**, 012009 (2023)
19. G. Vienne, Y. Xu, C. Jakobsen, H.J. Deyerl, J. Jensen, T. Sorensen, T. Hansen, Y. Huang, M. Terrel, R. Lee, N. Mortensen, J. Broeng, H. Simonsen, A. Bjarklev, A. Yariv, Ultra-large bandwidth hollow guiding in all silica Bragg fibers with nano supports. Opt. Express **12**, 3500–3508 (2004)
20. R.K. Chourasia, V. Singh, Estimation of photonic band gap in the hollow core cylindrical multilayer structure. Superlattice. Microst. **116**, 91–199 (2018)
21. R.K. Chourasia, N.K. Chourasia, N. Bihari, *Optical properties of hollow-core Bragg Fiber waveguides*, vol 23 (Benthem Science, 2023), pp. 214–236
22. D.R. Smith, R. Dalichaouch, N. Kroll, S. Schultz, S.L. McCall, P.M. Platzman, Photonic band structure without and with a defect in one-dimensional photonic crystals. J. Opt. Soc. Am. B Opt. Phys. **10**, 314–321 (1993)
23. L. González-García, S. Colodrero, H. Míguez, A.R. González-Elipe, Single-step fabrication process of 1-D photonic crystals coupled to nanocolumnar TiO2 layers to improve DSC efficiency. Opt. Express **23**, A1642–A1650 (2015)
24. Y. Gao, N. Guo, B. Gauvreau, M. Rajabian, O. Skorobogata, E. Pone, O. Zabeida, L. Martinu, C. Dubois, M. Skorobogatiy, Consecutive solvent evaporation and co-rolling techniques for polymer multilayer hollow fiber perform fabrication. J. Mater. Res. **21**, 2246–2254 (2006)
25. R.K. Chourasia, V. Singh, *Comparative analysis of the dispersion characteristic of Bragg fibers having different refractive index contrast*, in 13th International Conference on Fiber Optics and Photonics, OSA Technical Digest (online) (Optica Publishing Group, 2016), paper Th3A)
26. R.K. Chourasia et al., Performance study of a liquid-core Bragg fiber sensor in the presence of a defect layer. Opto-Electron. Rev. **25**(3), 215–221 (2017)

27. R.K. Chourasia, V. Singh, *Cylindrical Bragg waveguide based temperature sensors*, 2016 IEEE Uttar Pradesh Section International Conference on Electrical, Computer and Electronics Engineering (UPCON), Varanasi, 2016, pp. 87–90 (2016)
28. M.A. Kaliteevski, R.A. Abram, V.V. Nikolaev, G.S. Sokolovski, Bragg reflectors for cylindrical waves. J. Mod. Opt. **46**, 875–890 (1999)
29. A.H. Cherin, *An Introduction to Optical Fibers* (Mc Graw-Hill, Tokyo, 1983)
30. M. Born, E. Wolf, *Principles of Optics* (Cambridge University Press, Cambridge/London, 1999)
31. K. Kuriki, O. Shapira, S.D. Hart, G. Benoit, Y. Kuriki, J.F. Viens, M. Bayindir, J.D. Joannopoulos, Y. Fink, Hollow multilayer photonic bandgap fibers for NIR applications. Opt. Express **12**, 1510–1517 (2004)

Chapter 6
Multifunctional Bragg Fibers: To See, Hear, Sense, and Communicate Simultaneously

Abstract In this chapter, an intelligent and very fruitful application (see, hear, sense, and communicate) of Bragg fibers (BFs) designed through multifunctional, multilayered multi-materials will be discussed in detail. As we are aware, almost all electrical and optoelectronic devices need a complex arrangement of conducting, semiconducting, and insulating materials in precise shapes with interfaces that minimize scattering and feature dimensions. Several wafer-based processing methods have been created to meet these demands. However, these methods are limited by the size and shape of the wafer, as well as the intricate nature of the sequential, high-precision processing stages. On the other hand, the process of creating optical fibers by pulling them from a prepared rod is less complex and results in longer, more consistent fibers. Recently, a novel class of fibers consisting of conductors, semiconductors, and insulators has evolved. These fibers possess the fundamental characteristics of traditional electrical and optoelectronic devices but are manufactured using normal preform-based fiber-processing techniques, resulting in the production of functional fiber devices that span many kilometers in length. This study investigates two distinct methods for achieving complex functionalities: firstly, the incorporation of multiple functional components into a single fiber at the microscale level, and secondly, the construction of large-scale two- and three-dimensional geometric structures using multiple fibers. When utilized in conjunction, these two methodologies facilitate the development of versatile fabric systems.

6.1 Multi-Material Multifunctional Bragg Fiber Fabrication

The initial stage in the production of these versatile, layered materials involves creating a cylindrical object known as a preform. This preform has the same shape and composition as the final fiber, but is larger in size and shorter in length. The preform is subjected to thermal drawing to produce multi-material composite fibers [1–3]. These fibers are composed of at least two materials with distinct optical and electrical properties. The process preserves the original geometry of the preform while elongating the fibers and decreasing their cross-sectional dimensions. The

R. K. Chourasia, A. Katti, *Bragg Fibers*,
https://doi.org/10.1007/978-3-031-65164-9_6

crucial aspect of this procedure is to identify materials that can be pulled together and have the ability to preserve the shape of the preform in both the fiber and avoid any disruptions in the flow of liquid along the length and across the cross-section of the preform. In order to accomplish this goal, the utilization of viscous forces is frequently employed to counteract the capillary breakup mechanisms induced by interface energy [4–8]. One of the fiber materials must be able to withstand the drawing stress while also being able to deform continuously and controllably. Therefore, at least one component should be amorphous and resistant to devitrification. This will enable the fiber to be drawn at appropriate speeds in a furnace-tower process while maintaining its structural consistency. For proper functioning, it is necessary that all materials have a viscosity of less than 10^7 poise and are at the same temperature. Additionally, these materials should demonstrate strong adhesion and wetting properties in both their viscous and solid stages, without experiencing any cracking even under rapid thermal cooling conditions. Omnidirectionally reflecting structures impose more limitations on the characteristics of materials. In order to meet the requirement of omnidirectional reflectivity [9, 10], it is necessary for the two materials to have a high index contrast. Additionally, the materials should demonstrate little optical absorption across a wide range of wavelengths, ensuring that the absorption length is much greater than the evanescent decay length.

Our material selection has mostly concentrated on polymeric thermoplastics along with chalcogenide glasses having a high glass-transition temperature due to their continuous viscosity-temperature dependence and suitability for high-speed drawing. Some important examples of glass compositions are As_2Se_3, As_2S_3, $As_{40}Se_{50}Te_{10}Sn_5$, and $Ge_{15}As_{25}Se_{15}Te_{45}$. Thermoplastic polymers generally become soft at lower temperatures compared to normal chalcogenide glasses. Nevertheless, a small number of potential alternatives with thermal characteristics that are somewhat similar to chalcogenides have been discovered. Some examples of these polymers are polyether-sulphone (PES), poly-sulphone (PSu), and polyether-imide (PEI). It must be simultaneously drawn with glasses and polymers in order to create optoelectronic fiber devices. For the thermal-drawing method, only metals or alloys with low melting points are appropriate, as they need to have a melting temperature below the drawing temperature. Various metals, including Sn, In, Bi, and eutectics of Au, Bi, and Sn, have been integrated into fiber devices.

In order to provide a summary of the process of creating fibers, we have shown the process in Fig. 6.1. We examine a combined device that includes a cylindrical mirror structure that reflects light in all directions and an optoelectronic device made of metal, semiconductor, and insulator materials. The process of creating the fiber device starts by manufacturing a chalcogenide glass rod using conventional sealed-ampoule techniques [11, 12]. A polymer tube with a hollow center is created, with an inner diameter that precisely matches the outer diameter of the glass rod. The thickness of the tube is exactly the same as that of thin metallic ribbons. Subsequently, the glass rod is inserted into the tube, while the electrodes are placed into specially carved niches in the tube. Lastly, a shielding polymer cladding is wrapped over the entire construction. The metal electrodes are fully encapsulated between the polymer and the glass rod, ensuring no leakage occurs when it liquefies during the

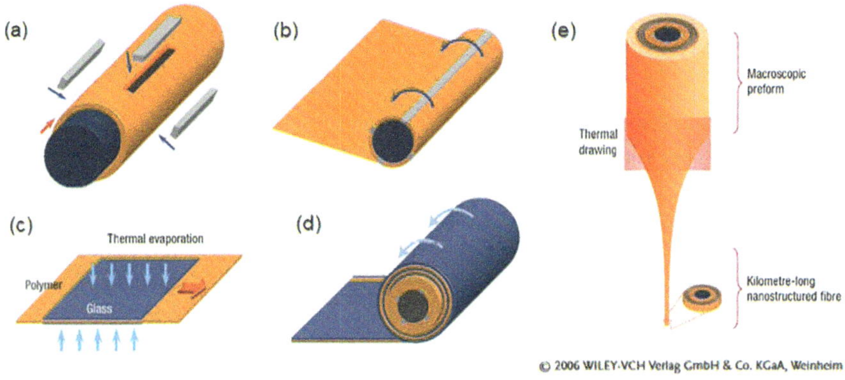

Fig. 6.1 Preform-based fabrication of integrated fiber devices. (**a**) A chalcogenide semiconducting glass rod is assembled with an insulating polymer shell and four metal electrodes, and (**b**) a polymer sheet is rolled around the structure to form a protective cladding. (**c**) The high-index chalcogenide glass is evaporated on both sides of a low-index thin polymer film before (**d**) being rolled around the cylinder prepared in (**a**, **b**). A polymer layer is wrapped around the coated film for protection. (**e**) The preform is consolidated in a vacuum oven and is thermally drawn to mesoscopic-scale fibers. The cross-section of the resulting fibers retains the same structure and relative sizes of the components at the preform level. (Reprinted with permission from [13])

drawing process. The process of creating hollow multilayer structures in fibers involves the heat evaporation of a high-index chalcogenide glass on both surfaces of a low-index thin polymer film that is not attached to any substrate. In order to produce a hollow fiber, the film is tightly wound around a silicate glass tube and then compressed by subjecting it to heat in a vacuum oven. The silicate tube is subsequently extracted from the preform core through the process of etching with hydrofluoric acid. When the structure is positioned on the outer surface of the fiber, there is no requirement for a quartz tube. Alternatively, the film is directly wrapped around a polymer cylinder, with a thin layer of protective polymer encompassing it. The preform depicted in Fig. 6.1 is made by combining both processes, wherein an exterior multilayer structure encloses the optoelectronic device. The preform is further compressed under vacuum conditions at an elevated temperature (usually around 10^{-3} torr and 260 °C). The fiber preform is thermally stretched into long lengths of fiber using the tower draw method, which is widely used in the fiber-optic sector. During the drawing process, the mirror layers are thinned by a factor of around 20–100, and the precise placements of the photonic band gaps (PBGs) are identified by monitoring the outside diameter of the fiber using a laser micrometer. The final outcome of this manufacturing process is the production of many meters of consistent fibers.

6.2 Omnidirectional Transmission and EM-Wave Guidance

Despite the long-standing knowledge and appeal of metallic mirror reflectors, their application in optics has been restricted due to significant optical losses. Conversely, multilayer interference filters have minimal optical losses but display significant reflection within a restricted angular range. A recent discovery has revealed the possibility of designing a finite one-dimensional periodic structure (as shown in Fig. 6.2a that combines the benefits of both and avoids the disadvantages, resulting in a mirror with low loss 360° reflectivity. To do this, certain requirements must be fulfilled about the refractive index contrast n_H/n_L (where n_H and n_L represent the refractive indices of the layers) and the ratio of the lower index to the ambient n_L/n_A (where $n_A = 1$ in this case). These conditions will result in a photonic band diagram similar to the one depicted in Fig. 6.2b. The picture displays the propagating Bloch states in brown color, while the prohibited states are represented in white. The diagram is plotted on a frequency ω versus wave vector β graph for both polarizations (TE and TM). The incident light from the surrounding environment is restricted to the part of the figure that lies above the light lines (where $\beta = \omega/c$). Light with frequencies within the gray range cannot interact with any propagating states in the structure and will always be reflected, regardless of the angle of incidence (ranging from 0° to 90°) or polarization. The omnidirectional reflector can be placed on the inside of a hollow-core cylindrical fiber to steer light along the axis [14] (Fig. 6.2c). The translation symmetry in the direction parallel to the layers is maintained in a finite cylindrical structure, while cylindrical waves are substituted for plane waves using Bessel functions [15]. The significance of the band diagram in this context lies in the anticipation that light will be concentrated within the core within a frequency range that aligns with the bandgap of a planar structure possessing identical structural parameters. The structure's boundary conditions permit the presence of discrete propagating modes within the bandgap, which are represented by the brown lines in Fig. 6.2d. The inset of Fig. 6.2d illustrates the vector field distributions of three modes of specific interest, namely HE_{11}, TE_{01}, and TM_{01} [16, 17]. Despite the existence of predictions regarding this fiber transmission mechanism as early as the 1970s, efforts to implement this structure were primarily hindered by two factors: (1) the periodicity of the multilayers must be maintained at a fraction of a wavelength across lengthy fiber lengths; and (2) the high-index contrast required between the layers typically necessitates significantly different thermomechanical properties, which eliminates conventional thermal drawing as a fabrication method. However, we have seen that the development of such a fiber is seen to be feasible despite the ostensibly insurmountable obstacles.

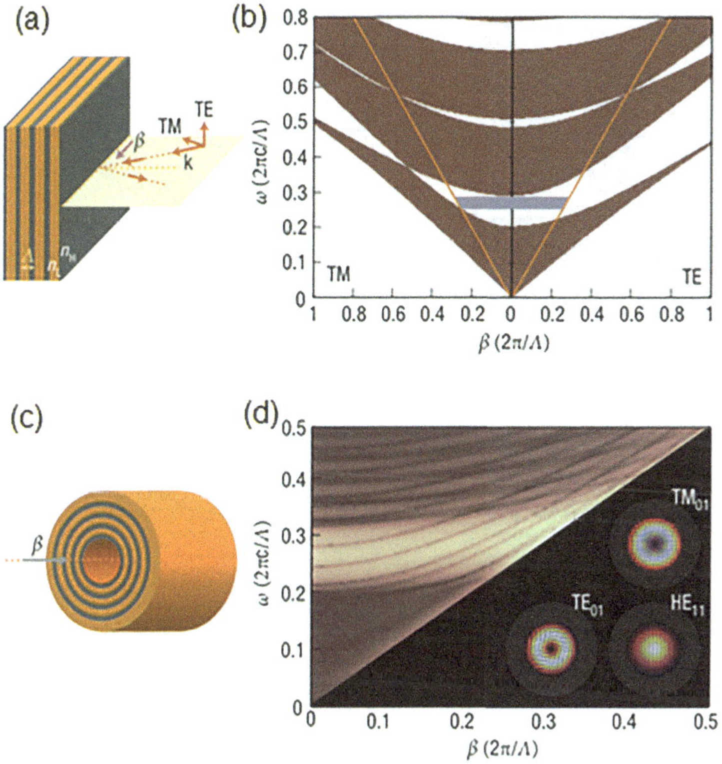

Fig. 6.2 Omnidirectional reflection and light guidance. (**a**) A 1D planar dielectric multilayer structure composed of alternating layers of indices n_L and n_H ($n_H > n_L$) of period Λ; β is the component of the wave vector *k* parallel to the structure. The direction of the electric field vectors for the transverse electric (TE) and transverse magnetic (TM) modes are shown. (**b**) The TE and TM band diagram of the structure in a. The two orange lines are the light lines in the ambient medium (ω/β = c, assuming the surrounding medium has $n = 1$). Propagating Bloch states are shown in brown whereas forbidden states are in white. (**c**) A hollow-core fiber lined with a cylindrical omnidirectional reflecting structure guides light along the axis (β is the axial wave number). (**d**) The band diagram for the fiber in c, where dark brown corresponds to allowed photonic states and white corresponds to forbidden states. Similar parameters to those for the structure in a were used, and the location of the PBG is similar to that in b for the planar structure. Note that allowed modes are now introduced into the PBG. The inset shows the vector field distribution of three low-order modes: TE_{01}, TM_{01} (both having angular momentum 0), and HE_{11} (hybrid mode, having angular momentum 1). (Reproduced with permission from [18])

6.3 Metal-Semiconductor-Insulator Bragg Fibers

In this section, we present the process of creating fibers that possess consistent electrical and optoelectronic capabilities across their whole length. This is achieved through the use of the preform-based fiber-drawing approach. The device is initially created as a cylindrical preform on a large scale, and then it is shrunk to the necessary

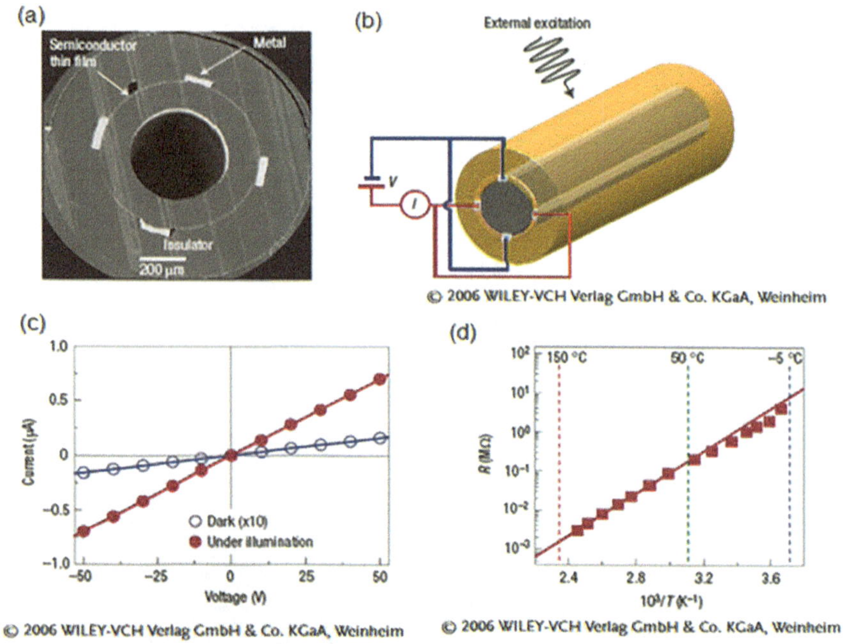

Fig. 6.3 Metal–semiconductor–insulator fiber devices. (**a**) SEM micrograph of a thin-film fiber device (the semiconductor is As_2Se_3, the insulator polymer is PES, and the metal is Sn). (**b**) The electrical connection of the four metal electrodes at the periphery of the fiber to an external electrical circuit. (**c**) The current–voltage characteristic curve of a photosensitive solid-core fiber device (980 μm outer diameter, 15 cm long). The conductivity increases upon illumination (20 mW, white light) when compared with dark conditions. (**d**) The resistance of a thermally sensitive solid-core fiber device (1150 μm outer diameter, 9 cm long) as a function of temperature. [(**a**) Reprinted from [18]. (**b–d**) reprinted with permission from [13]]

size via thermal drawing [19]. Figure 6.3a displays the cross sections of km-length functioning mesoscopic-scale device fibers, which are the outcome of the experiment. It is possible that the fundamental elements of contemporary electronic and optoelectronic devices, such as junctions and transistors, could be integrated into fibers created using this straightforward and inexpensive technique. This could be done on a larger scale than what is achievable using conventional electronics. When external light hits the amorphous semiconductor in the fiber core, it creates free charge carriers. The metal electrodes, which establish a connection with the core throughout the length of the fiber, are linked to an external circuit as shown in Fig. 6.3b. The fiber experiences a modification in electrical conductivity when exposed to external illumination, as seen by the alteration in the slope of the current-voltage curve when compared to dark conditions (Fig. 6.3c). Replacing the large semiconducting glass core with a thin-film glass layer (Fig. 6.3a) significantly enhances photosensitivity (by nearly two orders of magnitude) by removing the dark current generated by the non-penetrated volume of the core. This substitution also indicates broader potential applications for fibers. Figure 6.3e shows the temperature dependence of the resistance of a fiber with a solid core made of a thermally sensitive glass.

6.4 Bragg Optical Resonators

Lining a hollow-fiber core with an omnidirectional reflector confines light and guides it along the fiber axis. However, placing the reflector on the outer surface of the fiber will reflect light impinging on it externally with high efficiency [2]. The fiber structure, where an optical cavity defect is inserted into an exterior multilayer dielectric omnidirectional reflector [20, 21], leads to a polymer fiber that is at the same time a high-efficiency, omnidirectional reflecting mirror. These fibers may be incorporated into woven fabrics for precise spectral-identity verification or as radiation barriers, and may also be used as cost-effective tunable optical filters. The usefulness of this design has been further highlighted by producing cavity resonators in the external 1D photonic crystal, resulting in resonant tunneling at a frequency inside the PBG. Moreover, this resonant wavelength may be tuned dynamically. Using mechanical tuning, by applying axial stress to the fiber ends in the elastic regime, the applied strain leads to a linear shift of the Fabry-Pérot resonant mode within the bandgap [20].

6.5 Self-Monitoring Bragg Fibers

Hollow-core PBG fibers are potentially important for transmitting high optical power with applications in industrial and medical fields. As the power carried in such fibers increases, the potential adverse effects of the inadvertent release of this energy increase accordingly. Thus, predicting imminent failure is advantageous.

A fiber that is capable of sensing imminent failure is achieved using the structure shown in Fig. 6.4a, where a hollow-core PBG transmission fiber, designed to transmit a high-power CO_2 laser beam, is combined with a thin-film heat-sensing metal–semiconductor–insulator device [22] whose conductivity increases with increase in temperature. As the fiber conductivity depends exponentially on the peak temperature along its length, the different types of energy dissipation mechanisms may be distinguished, even when the amount of energy dissipated remains fixed. Figure 6.4b shows a thermal infrared image of a fiber section where a point defect was intentionally burnt into the fiber while the laser beam propagated through the fiber. The energy released at the localized defect raises the temperature of the fiber significantly in the vicinity of the defect (as detected by the camera), decreases the resistance of the semiconductor layer, and produces a large current in the MSM device (red circle in Fig. 6.4c). When the same energy is dissipated in a straight

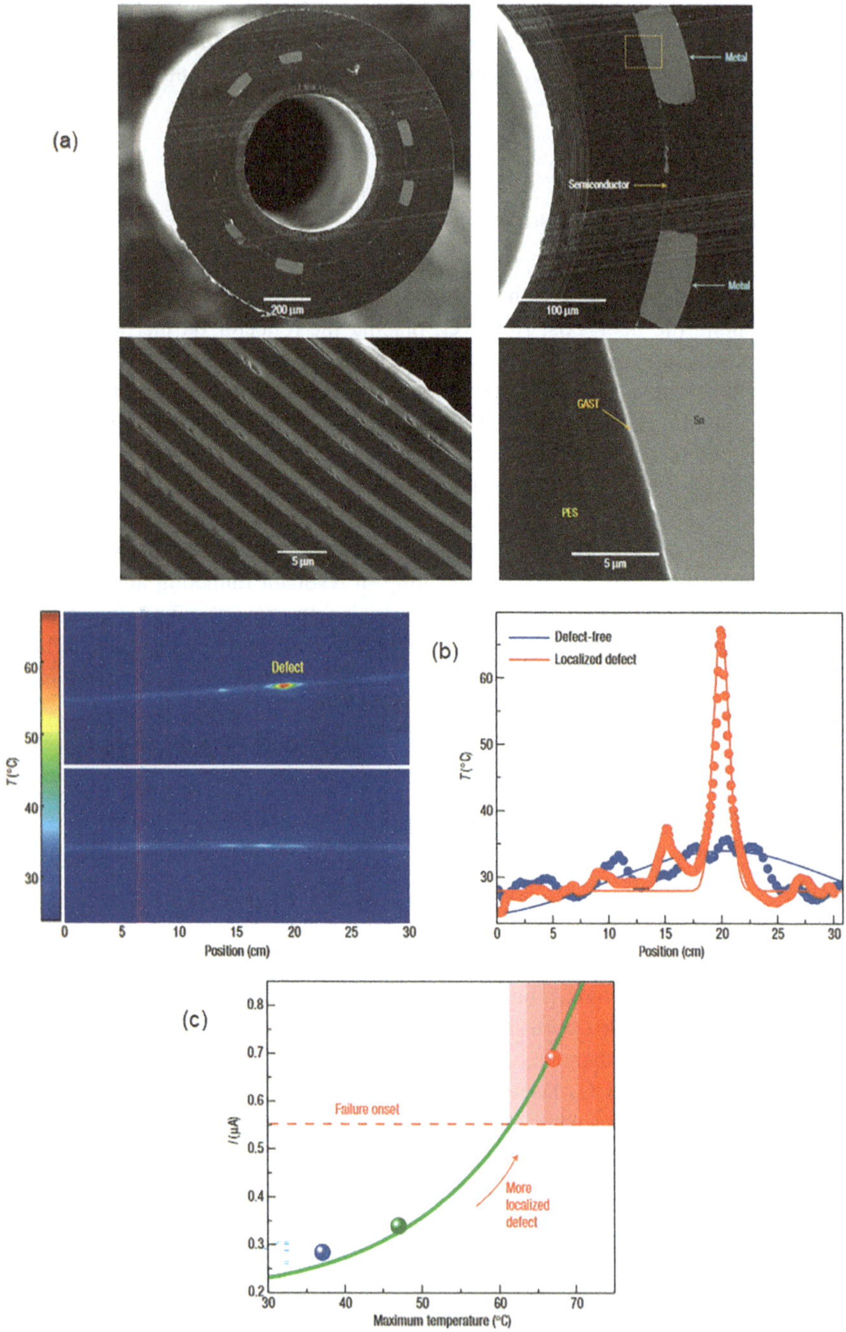

Fig. 6.4 Self-monitoring hollow-core fibers. (**a**) The top-left figure shows the SEM micrograph of a self-monitoring-fiber cross-section highlighting the different structures integrated into the device. The inset figures on the top-right and bottom-right show the MSM junction which is sensitive to thermal perturbations. GAST is gallium–arsenide–selenium–telluride. The lower left inset shows

defect-free fiber of same length, the peak temperature along the fiber is lower, and the measured current is also lower (blue circle in Fig. 6.4c). When the fiber was bent (bend radius 10 cm) and the same power dissipated, a peak temperature and measured current (green circle in Fig. 6.4c) that are intermediate between the defect-free and localized-defect cases were measured. This makes possible the placement of a threshold (*I* in Fig. 6.4c) for safe operation of the fiber—exceeding this threshold is a precursor to failure.

6.6 Narrowband Bragg Photodetectors

As the spectral range of photodetection of amorphous chalcogenide glasses is typically very wide, they are not useful as narrow-band photodetectors. Nevertheless, by combining the 1D photodetecting fibers with external omnidirectional reflecting structures, the penetration of a very narrow band of wavelengths to the photoconductive core can be engineered, resulting in a 1D narrowband photodetector [3]. This is done by placing a microcavity in the mirror structure to allow the desired wavelength (and desired bandwidth) to traverse the structure while reflecting all other wavelengths. Figure 6.5a shows this integrated fiber device highlighting the different structures included in it. Both the external reflectivity of the fiber (Fig. 6.5b) and the fiber photocurrent (Fig. 6.5c) were measured while scanning the incident wavelength. The functionality of this integrated device is found to be correctly correlated with the structure as shown by the shift of the resonant cavity defect wavelength with a change in the outer diameter of the device.

6.7 Surface-Emitting Bragg Fiber LASERs

All fiber lasers produced to date emit light along the fiber axis with a spot size dictated by the core radius and a nearly planar wavefront. Reflection from the end facets of the fiber provides the cavity required for lasing. There are many applications where it is beneficial to deliver laser light in the radial direction over an extended area, such as medical applications ranging from photodynamic therapy

Fig. 6.4 (continued) the multilayer structure lining the hollow core designed to guide a high-power CO_2 laser at 10.6 μm (see Fig. 6.1). (**b**) Thermal photographs of a fiber containing a single localized defect (upper panel) and a defect-free fiber (lower panel) captured by the infrared camera. Both fibers carry identical CO_2 laser energy. Temperature distributions taken along one-dimensional sections of the two fiber are shown. A defect was intentionally created in one of the fiber at the location of the peak in the measured temperature profile (fitted to a Gaussian profile). (**c**) Measurement of the MSM fiber device current at fixed dissipated power for three fibers: defect-free straight fiber (blue circle), a defect-free bent fiber (green circle), and a single-defect-containing straight fiber (red circle). (Figures reprinted with permission from [22])

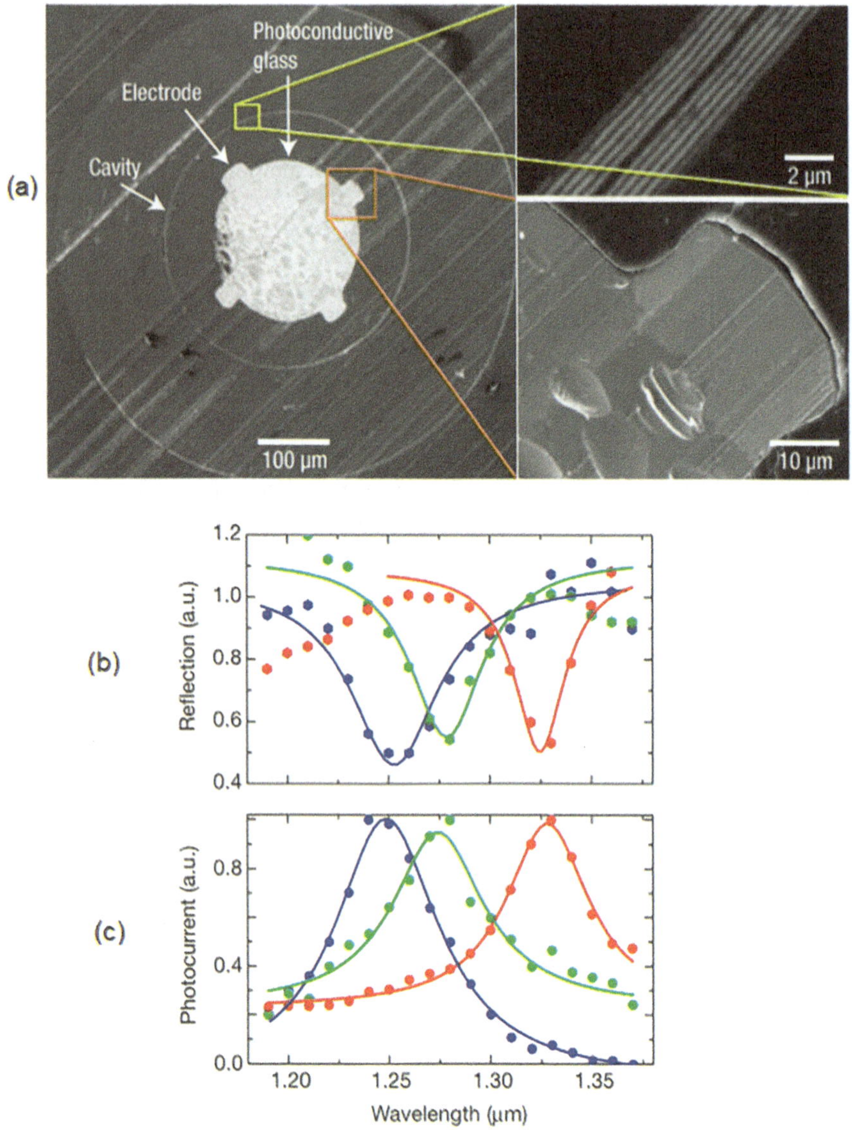

Fig. 6.5 Narrow-band photodetecting fibers. (**a**) SEM micrograph of a narrow-band photodetecting fiber cross-section highlighting the different structures integrated into the device. The inset on the lower right shows the metal–semiconductor interface and the upper right inset shows the external multilayer structure containing a low-index cavity. (**b**) Measured external reflectivity of three fibers with different outer diameters showing that the location of the resonance wavelength shifts proportionally to the diameter. (**c**) The photocurrent for the fibers in (**b**) is measured simultaneously with the reflectivity showing that the peak photocurrent occurs at the resonance wavelengths. (Figure reprinted with permission from [3])

[23] to in vivo molecular imaging [24], as well as textile fabric displays [25]. Omnidirectional reflection from a multilayer annular mirror lining a hollow-core fiber results in a radial lasing configuration [26]. The natural curvature of the PBG edges towards higher frequencies with an increase in wave number (see Fig. 6.2b) enables simultaneous optical confinement of two frequencies at distinct wave numbers. An optically pumped gain medium is inserted in the fiber core (in this specific case a fluorescent organic dye incorporated into a copolymer matrix) such that the emission spectrum of the medium lies within the omnidirectional bandgap at $\beta = 0$ (normal incidence to the structure). The higher-frequency optical pump is guided along the fiber axis (high wave number), whereas the lower-frequency laser emission is confined in the transverse direction (low wave number). This results in laser emission in the radial direction from the fiber surface at the location corresponding to the placement of the gain medium. The laser light emitted by this fiber laser has several unique features. The beam is emitted in a radial direction from an extended surface area. The optical wavefront is azimuthally anisotropic, having a dipole-like radiation pattern for a linearly polarized pump. Furthermore, the direction of laser emission is determined by the pump polarization: rotating the linear pump polarization results in rotating the emission direction. Lasers with emission lines throughout the visible and NIR may be constructed using a variety of dyes and scaling of the fiber bandgaps.

6.8 Bragg Fiber Web Networks

In the preceding part, we discussed a few examples of incorporating different structures into a single fiber in order to achieve a more advanced level of functioning. The insertion of more than one fiber into large-area arrays, which include functions that are not inherent in the single fibers, makes it possible to achieve an additional degree of freedom in terms of geometry. As an illustration, a single photosensitive (or thermally sensitive) fiber is unable to pinpoint the precise location of a point of excitation along its length. However, by arranging a multitude of similar fibers over a grid, it is possible to localize a point of optical (or thermal) excitation across the whole area of the grid. The detection of a point of optical or thermal excitation on a grid with dimensions of $N \times N$ typically necessitates the utilization of a number of N_2 point detectors. However, an array of linear 1D detectors can achieve the same capability with only the number of N detectors [22, 24]. This is in contrast to the conventional optical and thermal imaging systems described in [27, 28].

Figure 6.6a illustrates an example of such an array. In this particular instance, thermally sensitive fibers, which possess the structure seen in Fig. 6.3a, are woven into a fabric and then put on a curved surface. This results in the formation of an 8×8 array with a spacing of 1 centimeter [26]. This sample highlights a number of distinctive characteristics that are associated with the fibers. The fact that they are both flexible and lightweight makes it possible for them to be integrated into structures that already exist. Furthermore, because the fibers are manufactured in a

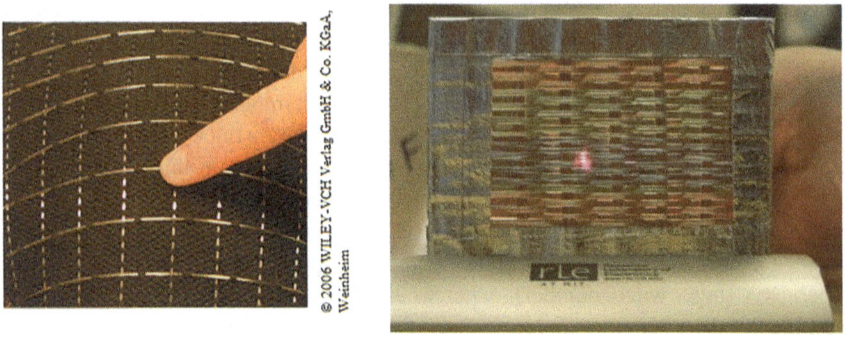

(a) (b)

Fig. 6.6 Arrays of optical and thermal fibers in both two and three dimensions. (**a**) A 2D array of thermally responsive fibers, arranged in an 8 by 8 grid with 1-cm gaps, and woven into a fabric that conforms to a curved surface. The lower panel displays a reconstruction of the heat distribution on the cloth following the contact of a finger. The reconstruction was performed by utilizing electrical-current measurements acquired from the fibers subsequent to the removal of the finger, and the data was graphed on a curved surface to enhance clarity. (**b**) The fabric is made by weaving together photosensitive fibers that are surrounded by omnidirectional reflectors containing resonant microcavities. (Refer to Fig. 6.5). This fabric has the capability to provide imaging and spectroscopic functions across a wide surface area. [Figure 6.6a reproduced with permission from [13]. Figure 6.6b reproduced with permission from [18]]

consistent manner in great lengths, the arrays that are constructed out of them have the potential to cover a wide area. Devices for photodetecting fibers, as was stated before (Fig. 6.5). Along its entire length, each fiber can detect light of a particular color wavelength. PBGs that are fundamental in the visible spectrum are present in the fibers, and they feature resonant cavities that enable just a restricted spectral band to get through to the core of the semiconducting glass crystal. The creation of a spectral image is accomplished by the collection of electrical impulses at the perimeter of the fibers in that array. The fibers have a tiny average diameter, which makes them more flexible and enables them to be woven into a fabric that is closely knit [22].

6.9 Future Scope and Vision

We anticipate that future research will concentrate on a number of important areas, including the discovery of novel materials, the enhancement of the quality of materials and interfaces, the development of new designs for fiber-devices, and the enhancement of device density through the decrease of feature sizes. This kind of processing has the potential to be compatible with a wide variety of polymer materials. These polymers include piezoelectric polymers, which are used for the purpose of fiber acoustic transduction, as well as conjugated polymers, which are used for light emission. In spite of the fact that we have, up to now, depended on inorganic chalcogenide glasses for optoelectronic functionality and simplicity of processing, it is anticipated that crystalline domains with higher electrical characteristics might be introduced into fiber devices by the induction of phase transitions [29, 30]. In turn, they may make it possible for electronic junction structures (such as p-n junctions, Schottky junctions, and so on) to be used for logic and optoelectronic operations. There are a great number of other qualities that make the glasses very handy. As an illustration, the Kerr nonlinearity measured by AsSe is about three orders of magnitude larger than that measured by silica glass [31–33]. It is therefore reasonable to anticipate the emergence of nonlinear optical effects that are established on these multi-material glass fibers. Some examples of these effects are the formation of supercontinuum [34, 35] and Raman amplification [36]. In addition, it possesses high magneto-optic (Verdet) coefficients [37], in addition to a multitude of other features, which immediately suggest a broad variety of small fiber-based devices. These fibers are continuing to expand their electronic capabilities, which will lead to the creation of some fascinating new gadgets in the near future. The incorporation of metallic components into these fibers makes it possible to create electrical and magnetic fields that are either static or variable over time. Specifically in biomedical applications, where discrimination between particles and pathogens can now be extended by directly probing not only their optical but also their electric and/or magnetic response, hollow-core fibers have been used for particle and atom guiding by radiation pressure [38]. The close proximity of controllable electric and magnetic fields enhances the capabilities of such systems, particularly in biomedical applications. There is a possibility that this line of study may eventually result in the creation of a "lab in a fiber," which would allow for the application of many diagnostic procedures in parallel or in cascade to particles that are propelled along the fibers. We consider the possibility of stacking and redrawing a number of different fiber-based devices into a single fiber in order to accomplish this integration (see Fig. 6.7). As to this moment, a single fiber structure has been responsible for the incorporation of eight different functional devices. Fabrics that are woven either entirely or partially from such interconnected fibers have the potential to offer a broad variety of real-time, nontraditional capabilities throughout the whole surface area of clothing. These functionalities can be fueled by electrical energy that is captured from the surrounding environment. The interaction between the characteristics of the materials and the integration of structures in these fibers, in conjunction with the production of fabric arrays, is only being started right now and has the potential to be a fascinating subject for both basic and practical research. The future development of transistors and fiber-integrated logic devices would bring about the possibility of genuinely multifunctional textiles as well as technological advancements that would bring about ever more advanced features.

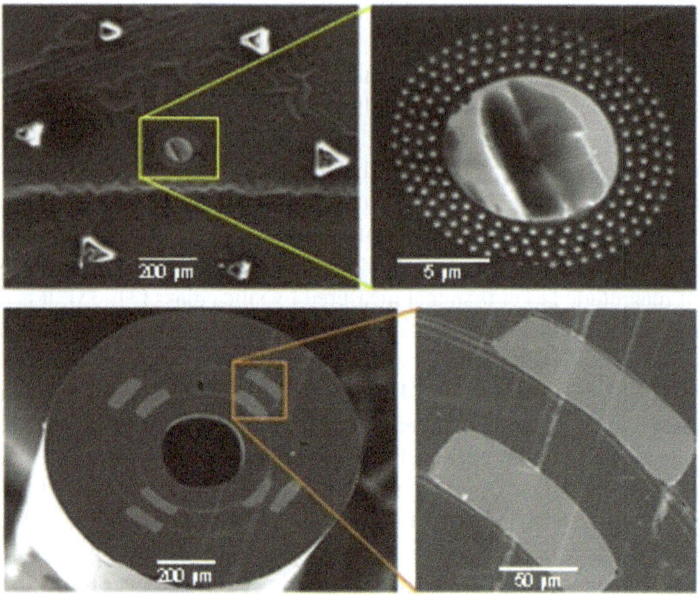

(a)

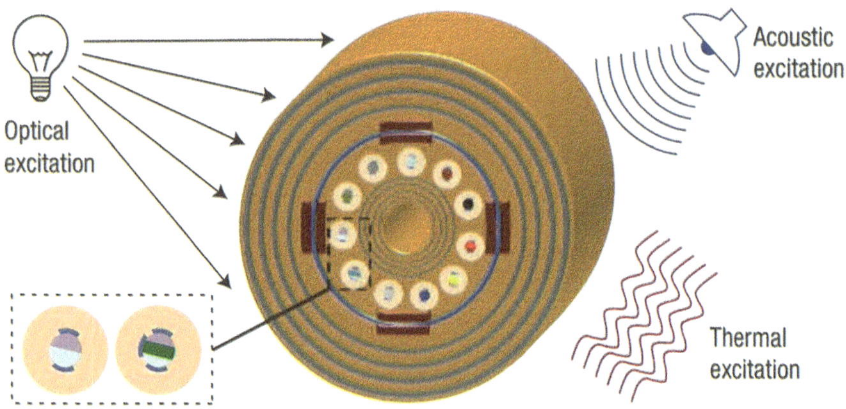

(b)

Fig. 6.7 Showcases bundles that have been formed by stacking and redrawing integrated fiber devices. (**a**) A collection of nanowires made of chalcogenide glass around a solid core made of extremely nonlinear chalcogenide glass, and two thin semiconductor film devices arranged in concentric layers inside the same fiber. (**b**) Envisioned future of combined fiber-device packages. A solitary fiber is composed of a hollow core that is coated with an omnidirectional reflector, which enables the transfer of optical power. The fiber is enclosed by an additional omnidirectional reflector, perhaps with many chambers, to spectrally filter incoming external radiation. The fiber comprises thin-film semiconducting devices, as well as many devices dispersed across the cross-section, with each device being responsive to a distinct environmental factor such as light, heat, and acoustic waves. The implementation of logical operations may also be achieved using basic semiconductor junctions, as illustrated by the two examples given in the inset. (Reproduced with permission from [18])

6.10 Conclusion

This chapter effectively addresses the design and production of multi-material, multifunctional BFs using preform-based methods. These fiber optic devices are made out of chalcogenide semiconducting glass and an insulating polymer bilayer. They also have four metal electrodes for the creation of new integrated devices. In addition, the article also explores the electrical and optoelectronic functions of metal–semiconductor–insulator BFs in order to achieve the necessary device appli cations. The concept of utilizing Bragg optical resonators with reflectors on the exterior surface of fiber resonators has also been explored for hollow Bragg fiber core bases. Self-monitoring beamforming systems are also potentially crucial for transferring large levels of optical power, with applications in the industrial and medical sectors. This chapter effectively addresses the use of narrow-band Bragg photodetectors, which employ amorphous chalcogenide glasses to detect a broad range of light wavelengths. This current book chapter additionally explores a highly sought-after and distinctive use of surface-emitting LASERs based on BFs. This chapter successfully explored the possibility of Bragg fiber web networking to fabricate the integrated Bragg fiber devices that play multitasking roles like seeing, hearing, sensing, and commuting with each other. The future scope and vision in Sect. 6.10 clearly mention the wide and fruitful scope of integrated Bragg fiber devices in the prime focus area like optoelectronic functionality, acoustic transduction, supercontinuum formation, and magneto-optic effect.

References

1. B. Temelkuran et al., Wavelength-scalable hollow optical fibres with large photonic bandgaps for CO2 laser transmission. Nature **420**, 650–653 (2002)
2. S.D. Hart et al., External reflection from omnidirectional dielectric mirror fibers. Science **296**, 510–513 (2002)
3. M. Bayindir et al., Metal-insulator-semiconductor optoelectronic fibres. Nature **431**, 826–829 (2004)
4. L. Rayleigh, On the stability of jets. Proc. Lond. Math. Soc. **10**, 4–13 (1878)
5. L. Rayleigh, On the capillary phenomena of jets. Proc. R. Soc. Lond. **29**, 71–97 (1879)
6. L. Rayleigh, On the instability of a cylinder of viscous liquid under capillary force. Philos. Mag. **34**, 145–154 (1892)
7. S. Tomotika, On the instability of a cylindrical thread of a viscous liquid surrounded by another viscous fluid. Proc. R. Soc. Lond. A **150**, 322–337 (1935)
8. J. Eggers, Nonlinear dynamics and breakup of free-surface flows. Rev. Mod. Phys. **69**, 865–929 (1997)
9. J.N. Winn, Omnidirectional reflection from a one-dimensional photonic crystal. Opt. Lett. **23**, 1573–1575 (1998)
10. Y. Fink et al., A dielectric omnidirectional reflector. Science **282**, 1679–1682 (1998)
11. A.R. Hilton, Optical properties of chalcogenide glasses. J. Non-Cryst. Solids **2**, 28–39 (1970)
12. Z.U. Borisova, *Glassy Semiconductors* (Plenum, New York, 1981)
13. M. Bayindir et al., Thermal-sensing fiber devices by multi-material codrawing. Adv. Mater. **18**, 845–849 (2006)

14. Y. Fink et al., Guiding optical light in air using an all-dielectric structure. J. Lightwave Technol. **17**, 2039–2041 (1999)
15. S.G. Johnson et al., Low-loss asymptotically single-mode propagation in large-core OmniGuide fibers. Opt. Express **9**, 748–779 (2001)
16. M. Ibanescu et al., Analysis of mode structure in hollow dielectric waveguide fibers. Phys. Rev. E **67**, 046608 (2003)
17. O. Shapira et al., Complete modal decomposition for optical waveguides. Phys. Rev. Lett. **94**, 143902 (2005)
18. A.F. Abouraddy et al., Towards multi-material multi-functional fibres that see, hear, sense and communicate. Nat. Mater. **6**(5), 336–347 (2007)
19. M. Bayindir et al., Fiber photodetectors co-drawn from conducting, semiconducting, and insulating materials. Opt. Photonics News **15**, 24 (2004)
20. G. Benoit et al., Static and dynamic properties of optical microcavities in photonic bandgap yarns. Adv. Mater. **15**, 2053–2056 (2003)
21. G. Benoit et al., Dynamic all-optical tuning of transverse resonant cavity modes in photonic bandgap fibers. Opt. Lett. **30**, 1620–1622 (2005)
22. M. Bayindir et al., Integrated fibres for self-monitored optical transport. Nat. Mater. **4**, 820–825 (2005)
23. R.M. Verdaasdonk, C.F.P. van Swol, Laser light delivery systems for medical applications. Phys. Med. Biol. **42**, 869–894 (1997)
24. V. Ntziachristos, C. Bremer, R. Weissleder, Fluorescence imaging with near-infrared light: New technological advances that enable in vivo molecular imaging. Eur. Radiol. **13**, 195–208 (2003)
25. V. Koncar, Optical fiber fabric displays. Opt. Photonics News **16**, 40–44 (2005)
26. O. Shapira et al., Surface-emitting fiber lasers. Opt. Express **14**, 3929–3935 (2006)
27. E.L. Dereniak, G.D. Boreman, *Infrared Detectors and Systems* (Wiley, New York, 1996)
28. A. Rogalski, History of infrared detectors. Opto-Electron. Rev. **20**(3), 279–308 (2012). https://doi.org/10.2478/s11772-012-0037-7
29. D.H. Ahn et al., A nonvolatile memory based on reversible phase changes between fcc and hcp. IEEE Electron. Dev. Lett. **26**, 286–288 (2005)
30. Z.M. Sun, J. Zhou, R. Ahuja, Structure of phase change materials for data storage. Phys. Rev. Lett. **96**, 055507 (2006)
31. G. Lenz et al., Large Kerr effect in bulk se-based chalcogenide glasses. Opt. Lett. **25**, 254–256 (2000)
32. S. Spalter et al., Strong self-phase modulation in planar chalcogenide glass waveguides. Opt. Lett. **27**, 363–365 (2002)
33. J.T. Gopinath et al., Third order nonlinearities in Ge-As-Se-based glasses for telecommunications applications. J. Appl. Phys. **96**, 6931–6933 (2004)
34. J.K. Ranka, R.S. Windeler, A.J. Stentz, Visible continuum generation in air-silica microstructure optical fibers with anomalous dispersion at 800 nm. Opt. Lett. **25**, 25–27 (2000)
35. T.A. Birks, W.J. Wadsworth, P.S. Russell, Supercontinuum generation in tapered fibers. Opt. Lett. **25**, 1415–1417 (2000)
36. R.E. Slusher et al., Large Raman gain and nonlinear phase shifts in high-purity As2Se3 chalcogenide fibers. J. Opt. Soc. Am. B **21**, 1146–1155 (2004)
37. Y.L. Ruan et al., Wavelength dispersion of Verdet constants in chalcogenide glasses for magneto-optical waveguide devices. Opt. Commun. **252**, 39–45 (2005)
38. M.J. Enn et al., Laser-guided atoms in hollow-core optical fibers. Phys. Rev. Lett. **75**, 3253–3256 (1995)

Chapter 7
Optofluidic Bragg Fiber Sensor Applications: Fuel Adulteration Sensor (Perceiving in Chemically Diverse Environments)

Abstract Acquiring a pleasant smell in environments with a high degree of chemical complexity has been a difficult task up until now. This chapter centers on enhancing and documenting the efficiency of optofluidic Bragg fibers (BFs) artificial noses. These noses are specifically engineered to measure the proportion of an impure pseudo-binary mixture of water-containing single-alcohol fuel, with the measurement relying on temperature. The objective entails utilizing mathematical models to accurately forecast the behavior of an optofluidic BFs sensor system that contains geometrical flaws. This is accomplished by introducing a lack of homogeneity in the central region of periodic cylindrical Bragg reflectors. The simulation of a multilayer concentric hollow-core BFs structure utilizes a cylindrical coordinate system. The simulation is conducted using Hankel functions (HFs) and the transfer matrix method (TMT). The fluctuation in refractive index (RI) of the contaminated pseudo-binary mono-alcohol fuel is linked to the temperature-dependent molar concentration. Multiple theories, such as the Dale-Gladstone and Lorentz-Lorenz models, are utilized to predict this correlation. The study noticed a sharp and well-defined peak in signal transmission, with a width of 0.1 nm at half of its highest intensity. This peak appeared in the photonic bandgap and showed a strong response to changes in the refractive index of the optofluidic core, particularly near a structural defect cavity. The proposed sensor exhibits a temperature-dependent maximum sensitivity of 1057.32 nm/RIU specifically for ethanol fuel, as opposed to methanol fuel, under different weather conditions. Furthermore, a comparison is drawn between a static temperature sensor based on surface plasmons. The proposed optofluidic BFs mono-alcohol fuel adulteration sensor device demonstrates enhanced sensing performance metrics, encompassing quality parameters and detection accuracy. This is attributed to the output signal's FWHM (full width at half maximum), which is notably brief at approximately 0.1 nm.

R. K. Chourasia, A. Katti, *Bragg Fibers*,
https://doi.org/10.1007/978-3-031-65164-9_7

7.1 Introduction

Adulteration refers to the deliberate addition of unauthorized or illegal foreign substances to a main fuel or similar substance, resulting in the product failing to meet its necessary standards and specifications. Adulterants are extraneous substances that modify and degrade the quality of crucial transportation fuels upon their addition. When there are discrepancies in the prices of things of the same quality or when customers are unable to distinguish between similar objects of different standards, unscrupulous individuals may attempt to unlawfully exploit the circumstances. Fuel adulteration and under-dispensing are significant forms of misconduct in the retail business. These strategies result in losses across several domains, such as engine malfunction and declining air quality. Fuel tax evasion has a direct impact on the revenue of the federal government. Consumer losses occur due to inadequate distribution of commodities. The significant energy capacity of biomass-derived fuels is widely acknowledged; yet, one of their primary benefits lies in their capacity to mitigate greenhouse gas emissions. These fuels offer a sustainable energy option that will gradually reduce dependence on nonrenewable resources such as petroleum, while also contributing to the establishment of a strategic reserve [1, 2]. Pure or hydrated mono-alcohols are a type of biomass fuel that is widely used in several industries, such as pharmaceuticals and cosmetics [3]. When it comes to the sources of energy, it is generally accepted that the availability of diesel, petrol, and other fossil fuels is limited. Mono-alcohol has the potential to function as an environmentally sustainable biofuel, replacing diesel, gasoline, and other nonrenewable fuel sources [4, 5]. Significant amounts of mono-alcohol are typically derived from sugarcane, switchgrass, corn, natural gas, municipal garbage, and lignite [6]. The promotion of internationally standardized, high-quality fuels can be effectively achieved by implementing the Worldwide Fuel Mandate [7] and following the expert committee's suggestions for the roadmap of ethanol blending in India by 2025 [8]. Additionally, the need for engine and vehicle durability optimization and the most environmentally benign engine and vehicle technology are incorporated into this strategy. Consumers will benefit from the implementation of the specific objectives, which will increase the efficiency of petroleum markets, stimulate the economy, and aid governments in the attainment of their public policy goals.

Today, mono-alcohol hydrocarbons like methanol and ethanol are used both as direct additives to gasoline and as feedstock for the additives that are synthesized from them. These hydrocarbons are regarded as sustainable. Physical-chemical properties of methanol and ethanol are comparable, including solubility in water, density, appearance, odor, and transparency [9]. The reduced expense of water relative to mono-alcohols, coupled with their similar properties, contributes to the susceptibility of hydrated mono-alcohol fuel to contamination. The maintenance of high-quality ethanol's effectiveness as both a fuel and a supplementary fuel additive source is directly proportional to its purity. The manufacturer's guidelines for the recommended ethanol quality required to attain optimal engine and vehicle performance are detailed in this document. The prescribed limits are applicable to a 100% pure anhydrous ethanol blendstock that is intended for combination with gasoline derived from petroleum to produce a blend with a maximum ethanol content of 10% by volume. E10 is a suitable mixture for implementation in vehicles that are outfitted with spark-ignition engines. Blends of fuel containing more than E10's percent ethanol [10] are intended for use exclusively in vehicles that are specifically engineered to operate on such fuel, such as "flex-fuel" automobiles. Furthermore, ethanol utilized in higher blends may be subject to additional restrictions. Ethanol has emerged as the prevailing biofuel on a global scale, surpassing 90 billion liters produced annually over the last 10 years. This type of energy matrix can be utilized in lieu of fossil fuels in Mexico, Europe, the United States, Brazil, Japan, India, and Argentina. Fuel efficacy is improved by the addition of anhydrous ethanol to gasoline, which acts as an octane booster [10]. Excessive water addition is the primary method of adulterating mono-alcohol fuel, as the blends are transparent and lack any discernible scent. It leads to a decrease in power and an elevated rate of fuel consumption. Furthermore, the molar concentration of mono-alcohol exhibits substantial variations in response to air temperature and other climatic conditions [11–14]. In light of the widespread use of all-season tires in Asia, including India, it is imperative to diligently observe the impurity content of hydrated mono-alcohol mixtures across varying weather conditions. The temperature influences and interconnects the refractive index (RI), viscosity, and density, which are all physical properties of this pseudo-binary combination [12–14]. Temperature fluctuations cause a variation in the molar concentration of hydrated mono-alcohols, which influences the factor involved. A multitude of techniques have been proposed to quantify the water content in mono-alcohol fuels. These include the following: infrared spectroscopy [15], surface plasmon sensor [21], direct-injection enthalpimetry [16], cyclic voltammetry [17], photothermal detector [18], webcam photometer [19], and ultrasonic propagation velocity [20]. These techniques demonstrate efficacy solely for mono-alcohols when subjected to static temperature conditions. To ensure adherence to the permissible thresholds of hydrated mono-alcohol contamination, it is crucial to devise a cost-efficient, uncomplicated, and effective method for identifying changes in molar concentrations that are influenced by temperature. Furthermore, this approach ought to improve the precision of identification.

In this chapter, we shall be expanding our analysis to include the impact of temperature changes caused by varying climatic and meteorological conditions, in contrast to previous studies done in static environments and at constant temperatures. The molar concentration of adulterants in hydrated mono-alcohol fuel is measured accordingly. As the sensing substrate for this task, we implemented the hollow-core optofluidic Bragg fiber adulteration sensor (BFAS) featuring a defect layer to identify mono-alcohol fuel adulteration. Based on our current understanding, this suggested research is completely novel and has been the first to successfully utilize BFAS in the context of detecting adulteration in mono-alcohols. The BFAS structure is built around a concentric circular layer arrangement with varying refractive indices. This arrangement forms the core foundation of the structure.

The construction allows for the transmission of EM waves through the empty core, which undergoes repeated reflections due to the periodic claddings. The BFAS structure must exhibit a photonic bandgap (PBG) under Bragg conditions [22–27]. The PBG range refers to a specific group of wavelengths that are prohibited from passing through the BFAS device configuration. The BFAS device arrangement has several promising applications in various fields such as doubly tuned multiwavelength LASER sources [28], biosensors [29–32], voltage adjustable multiplexers [33], intensity modulation, wavelength modulation [34], among others, and biofuel adulteration sensors [35–37]. This study examines the contrast of materials with high refractive indices (RI), including chalcogenide glass, arsenic tri-selenide, and polymer film polyether-imide (PEI). Additionally, the configuration of a multi-clad BFAS device is investigated. During the sensing procedure, the vacant area present in the BFAS device configuration will operate as an optofluidic flow cell specifically designed for hydrated mono-alcohols. To optimize the functionality of sensing devices, it is imperative that the output signal maintains a consistent strength and a narrow transmission bandwidth when utilized inline. To attain this characteristic in the current periodic structure, a PEI layer with a geometric imperfection or defect is incorporated into the cladding periodicity. As a consequence of this asymmetry and disruption of symmetry, a resonant transmission peak is produced in the PBG region. The transmission characteristics of the electromagnetic wave through the BFAS device configuration were examined in the cylindrical coordinate system using the TMT and mathematical high-frequency formalism.

7.2 Comprehensive Outline of Bragg Fiber Fuel Adulteration Sensor (BFAS) Device Setup

The geometrical flaw-laden optofluidic core BFAS device is depicted in Fig. 7.1a, which illustrates the cladding RI and thickness profile of a pseudo-binary combination of water and mono-alcohols. Circular symmetry characterizes the longitudinal axis (propagation axis) of the BFAS device configuration. The periodic repeating of bilayer high RI chalcogenide glassy substance Arsenic-tri-Selenide As_2Se_3 (n_H, d_H) and low R.I. polymeric layer poly-ether-imide (PEI) creates the BFAS device setup (n_L, d_L). In addition, the periodic arrangement's symmetry is disrupted by the introduction of a polymeric layer with a low refractive index termed polyether-imide (PEI). This layer acts as a geometrical defect layer and is characterized by (n_D, d_D). The quarter wave stack condition $n_H d_H = n_L d_L = n_D d_D = \lambda_0/4$ applied on the thicknesses ensures maximum intensity of transmission, where λ_0 is the BFAS device setup design wavelength. Furthermore, Fig. 7.1b illustrates the schematic arrangement for enhancing the temperature-dependent molar concentration of the pseudo-binary mixture of water and mono-alcohols by the utilization of this BFAS adulteration device setup. The electromagnetic waves emitted by the multi-wavelength broadband source are directed towards the optofluidic core BFAS device, which is

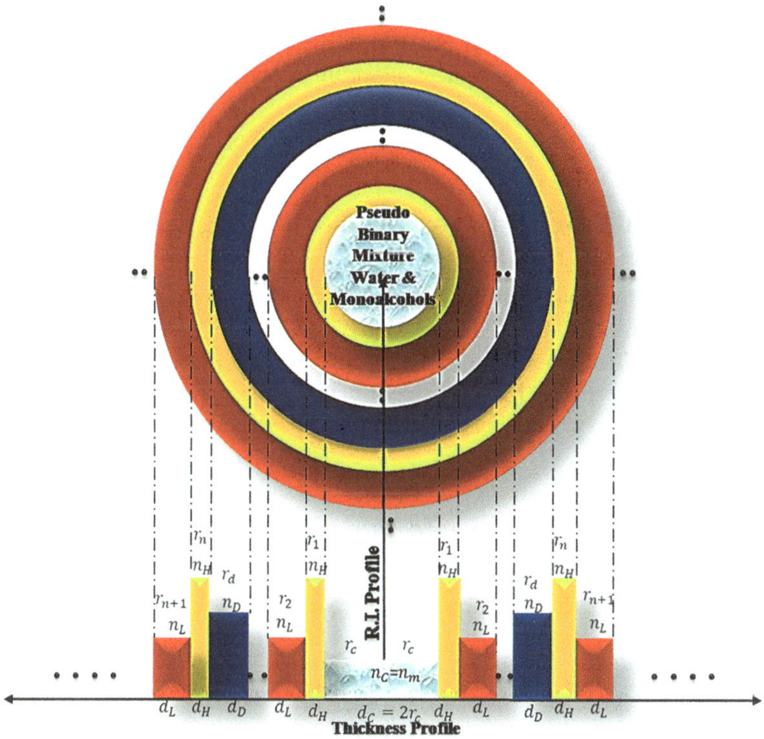

Fig. 7.1a Thickness and RI profile of BFAS device setup. The geometrical defect cavity is shown and it is noticeable that the optofluidic core is filled with hydrated mono-alcohols [37]. (Reprinted from Heliyon, 8, Nitesh K. Chourasia, Narendra Bihari, Ritesh Kumar Chourasia, An optofluidic Bragg fiber sensor for estimating adulterants in a temperature-dependent molar fraction of hydrated mono-alcohol fuels, e10532., Copyright 2022, with permission from Elsevier)

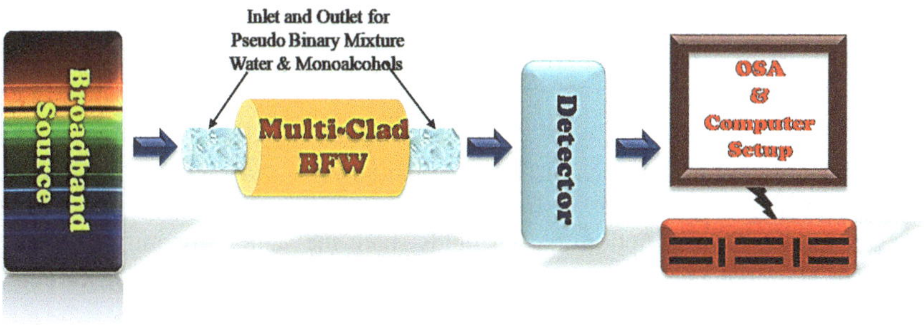

Fig. 7.1b BFAS optofluidic device configuration schematic for optimizing the temperature-sensitive molar concentration of adulterant in hydrated mono-alcohols [37]. (Reprinted from Heliyon, 8, Nitesh K. Chourasia, Narendra Bihari, Ritesh Kumar Chourasia, An optofluidic Bragg fiber sensor for estimating adulterants in a temperature-dependent molar fraction of hydrated mono-alcohol fuels, e10532., Copyright 2022, with permission from Elsevier)

filled with a pseudo-binary mixture. This arrangement is depicted in a simple schematic form. Subsequently, a photo-optic detector is employed to detect the transmitted waves. Finally, an optical spectrum analyser analyses the transmitted waves.

7.3 Optofluidic BFAS Device Setup Modeling

The structure of the optofluidic BFAS device has been modeled theoretically and mathematically using TMM (Transfer Matrix Method) and HF (High Frequency) methods in cylindrical coordinates. The subsequent text presents the RI profile of many layers comprising the BFAS device structure utilized in this specific chapter of the book:

$$\begin{gathered} n(r) = (n_c, 0 \leq r < r_c; \\ n_H, r_c \leq r < r_1; \\ n_L,\ r_1 \leq r < r_2; \\ n_H,\ r_{n-2} \leq r < r_{n-1}; \\ n_L, r_{n-1} \leq r < r_D; \\ n_D, r_D \leq r < r_{D'}; \\ n_H,\ r_{D'} \leq r < r_{n+1}; \\ n_L, r_{n+1} \leq r < r_{n+2}; \text{etc.} \end{gathered} \tag{7.1}$$

The cylindrical wave TMT [22, 23] can be utilized to construct the necessary BFAS device setup, obtaining the solution for both reflection and transmission. The Maxwell equations [38, 39] describe the behavior of a substrate with permeability "μ" and permittivity "ε";

$$\nabla \times E = -j\omega\mu H \tag{7.2}$$

$$\nabla \times H = j\omega\varepsilon E \tag{7.3}$$

The interconnection between the longitudinal and transverse components of the electric and magnetic fields is defined in the cylindrical coordinate system (r, z) [38, 39], under the assumption that all fields share the same temporal aspect exp $(j\omega t)$. The Helmholtz wave equations can be formulated in the cylindrical coordinate system as follows:

$$r\frac{\partial}{\partial r}\left(r\frac{\partial H_z}{\partial r}\right) - r^2\frac{1}{\varepsilon}\frac{\partial \varepsilon}{\partial r}\frac{\partial H_z}{\partial r} + \frac{\partial}{\partial \theta}\left(\frac{\partial H_z}{\partial \theta}\right) + \omega^2\mu\varepsilon r^2 H_z = 0 \text{ (TE–mode)} \tag{7.4}$$

$$r\frac{\partial}{\partial r}\left(r\frac{\partial E_z}{\partial r}\right)-r^2\frac{1}{\mu}\frac{\partial \mu}{\partial r}\frac{\partial E_z}{\partial r}+\frac{\partial}{\partial \theta}\left(\frac{\partial E_z}{\partial \theta}\right)+\omega^2\mu\varepsilon r^2 E_z=0\ \text{(TM–mode)} \quad (7.5)$$

The differential equations mentioned earlier are amenable to solution through the variable segregation procedure. In the cylindrical coordinates, the angular and radial components of the Maxwell field solutions are obtained [22, 39]. The nontrivial components (H_Z, E_θ, E_r) of the TE-mode solution can be written as,

$$H_Z(r,\theta)=[AJ_m(kr)+BY_m(kr)]e^{jm\theta} \quad (7.6a)$$

$$E_\theta=\frac{1}{-j\omega\varepsilon}\frac{\partial H_z(r,\theta)}{\partial r} \quad (7.6b)$$

$$E_r=\frac{m}{\omega\varepsilon}\frac{H_z(r,\theta)}{r} \quad (7.6c)$$

where A and B are arbitrary constants, J_m represents the Y_m Bessel and Neumann functions, $k=\left(\frac{\omega}{c}\right)n$ represents the medium's wave vector, and "n" denotes the medium's RI By applying Abeles' multilayer structure theory [22, 39], the elements of the transfer matrix $\widehat{T}$, that establish a correlation between the field vector and another point "r" from the initial position r_c are computed. These elements are defined as follows:

$$\widehat{T}=[T_{11}\ T_{12}\ T_{21}\ T_{22}] \quad (7.7)$$

$$T_{11}=\frac{\pi}{2}kr_c[Y'_m(kr_c)J_m(kr)-J'_m(kr_c)Y_m(kr)] \quad (7.8a)$$

$$T_{21}=j\frac{\pi}{2}kr_c p[Y'_m(kr_c)J'_m(kr)-J'_m(kr_c)Y'_m(kr)] \quad (7.8b)$$

$$T_{22}=\frac{\pi}{2}kr_c[J_m(kr_c)Y'_m(kr)-Y_m(kr_c)J'_m(kr)] \quad (7.8c)$$

$$T_{12}=-j\frac{\pi}{2}\frac{kr_c}{p}[J_m(kr_c)Y_m(kr)-Y_m(kr_c)J_m(kr)] \quad (7.8d)$$

where $p=\sqrt{\frac{\mu}{\varepsilon}}$ is the medium's intrinsic impedance. The calculation of the asymmetric defect cavity transfer matrix components must be performed using the techniques indicated above. As the consideration of both radii of two interfaces is necessary for the calculation of the transfer matrix elements, the nth transfer matrix $\widehat{T}_n$ can therefore be determined by substituting $r_c \rightarrow r_{n-1}, r \rightarrow r_n, k \rightarrow k_n, p \rightarrow p_n$. Additionally, as the setup of the BFAS device is periodic, we must take $n(r)=n_L$ if $n=$ even and $n(r)=n_H$ if $n=$ odd. We have total (2 N + 1) layers in the BFAS device setup with an asymmetric low R.I. defect cavity; thus, the overall transfer matrix is:

$$\left[V\left(r_f\right)\ U\left(r_f\right)\ \right] = T_{2N}\ldots\ldots\ldots\ldots T_2T_1[V(r_c)\ U(r_c)\ ] = \widehat{T}[V(r_c)\ U(r_c)\ ] \quad (7.9)$$

The solutions of the Helmholtz equation at positions r and V are represented as $U(r)$ and $V(r)$, respectively. Transfer matrices, which are equivalent to Fresnel's equation [38] in a planar geometry, are utilized to calculate reflection and propagation across multiple layers, as demonstrated previously. The calculation of the reflection coefficient using the BFAS device setup is as follows:

$$r_M = \frac{\left(T'_{21} - jp_0c_{m0}^{(2)}T'_{11}\right) + jp_fc_{mf}^{(2)}\left(T'_{22} - jp_0c_{m0}^{(2)}T'_{12}\right)}{\left(-T'_{21} + jp_0c_{m0}^{(1)}T'_{11}\right) + jp_fc_{mf}^{(2)}\left(-T'_{22} + jp_0c_{m0}^{(1)}T'_{12}\right)} \quad (7.10)$$

In Eq. (7.10), the matrix element T'_{11}, T'_{22}, T'_{12}, and T'_{21} are the inverse matrix element of matrix $\widehat{T}_m$ in Eq. (7.9) and

$$C_{ml}^{(1,2)} = \frac{H_m^{(1,2)\prime}(k_l r_l)}{H_m^{(1,2)}(k_l r_l)} \text{ with } l = 0, f \quad (7.11)$$

The first and second kind Hankel functions can be defined by $H_m^{(1)}$ and $H_m^{(2)}$ for the azimuthal field variation m. The reflectance can be measured as $R = |r_M|^2$, and the percentage transmittance for nonabsorbing media is calculated accordingly, $T = (1 - R) \times 100$ with values between 0% and 100%. The *TM*-mode formula can be attained basically by inserting $\mu \leftrightarrow \varepsilon, j \leftrightarrow -j$ in the *TE*-mode expression.

7.4 The Correlation Between the R.I. of Hydrated Mono-alcohols with Composition and Temperature

Monohydric alcohols, whether in their pure form or when mixed with water, are widely utilized in various applications, with our key focus in the research being on energy provision. The surface tension, density, refractive index, viscosity, and other physical properties of these pseudo-binary fuel resources provide valuable analytical data for the fuel industry [12–14]. Our suggested optofluidic BFAS device setup requires us to focus solely on two specific physical parameters: refractive index (RI) and density. Hence, other equations such as Lorentz-Lorenz, Dale-Gladstone, Eykman, Newton, and Kay have been developed to predict certain outcomes. However, the first two equations, Lorentz-Lorenz and Dale-Gladstone, have proven to be particularly effective in explaining theoretical predictions and achieving experimental agreement [13]. At ambient temperature (298.15 K) and atmospheric pressure, the refractive index (RI) and density (ρ) of a pseudo-binary mixture of mono-alcohols and water (hydrated mono-alcohols) were determined. The refractive indices (RIs) of these solutions have been examined in relation to temperature within the range of 278.15–323.15 K (which matches to the expected weather temperature range) and atmospheric pressure. The refractive indices (RIs) acquired through experimentation were compared to those estimated using the Lorentz-Lorenz and Gladstone-Dale relations in references [13, 14]. The results obtained from the use of the Gladstone-Dale equation were consistently comparable to the experimental observations in all instances [13] for mono-alcohols. The V^E, excess molar volumes and ΔR, molar volume fraction deviations of these combinations have been calculated, along with the disparities between the experimental and Gladstone-Dale refractive index values. When it comes to estimating the refractive index (RI) of pseudo-binary mixtures of liquids, the Lorentz-Lorenz relation is the most widely used theoretical approach [13],

$$\frac{n_L^2-1}{n_L^2+2}=\varnothing_a\frac{n_{Da}^2-1}{n_{Da}^2+2}+\varnothing_b\frac{n_{Db}^2-1}{n_{Db}^2+2}\ \text{(Lorentz–Lorenz)} \tag{7.12}$$

The Lorentz-Lorenz formula calculates the refractive index (R.I.) of a mixture, denoted as n_L. The molar percentage of hydrated mono-alcohols in the mixture, represented by $\varnothing_a$ and $\varnothing_b$, is determined using volume measurements based on mass and density. The individual refractive indices of the mono-alcohols and water are denoted as n_{Da} and n_{Db} respectively. An alternative method to compute the refractive indices (RIs) of pseudo-binary liquid systems is by the use of empirical equations. One such equation, proposed by Gladstone-Dale [13], has been developed for this purpose.

$$\frac{n_G-1}{\rho}=w_a\left(\frac{n_{Da}-1}{\rho_a}\right)+w_b\left(\frac{n_{Db}-1}{\rho_b}\right)\ \text{(Gladstone–Dale)} \tag{7.13}$$

The variables "n_G" and "ρ" represent the refractive index and experimental density of the mixture, respectively. w_a and w_b denote the mass fractions of mono-alcohols and water. ""n_{Da}" and "'n_{Db}" refer to the refractive indices of the individual mono-alcohols and water, respectively. It has been demonstrated by various scholars that the Lorentz-Lorenz and Gladstone-Dale relationships are valid when applied to a variety of different kinds of mixtures [12–14]. The excess molar volume, denoted as "V^E," has been calculated using the equation [13].

$$V^E=\frac{x_aM_a+x_bM_b}{\rho}-\frac{x_aM_a}{\rho_a}-\frac{x_bM_b}{\rho_b} \tag{7.14}$$

The symbol "ρ" represents the density of the mixture. "'ρ_a" and ""ρ_b" represent the densities of the individual components. "x_a" and "x_b" represent the molar fractions of the components. "M_a" and "M_b" represent the molar masses of the pure components.

In addition, the deviations in molar refraction, denoted as "ΔR", have been estimated using the following formulas in order to enhance the precision of the observations.

$$R_m = \left(\frac{n_D^2 - 1}{n_D^2 + 2}\right)\frac{x_a M_a + x_b M_b}{\rho} \tag{7.15}$$

$$\Delta R = R_m - (x_a R_{a+} x_b R_b) \tag{7.16}$$

where "n_D" represents the refractive index (R.I.) of the mixture. Conversely, researchers have also conducted computations to determine the disparities between empirically measured refractive indices (R.I.) denoted as "n_D" and those calculated using the Gladstone-Dale equation, denoted as n_G. This difference is represented as Δn_G, which is equal to the subtraction of "n_G" from "n_D."

7.5 Optofluidic BFAS Adulteration Sensor Device Setup Performance Parameters (P.P.)

This study assesses the sensitivity (S), detection accuracy (D.A.), and overall quality parameter (Q.P.) of the suggested optofluidic BFAS adulteration concentration sensor to evaluate its performance. The sensor's sensitivity is calculated by dividing the measured output by the change in the physical quantity being monitored. The sensitivity of the current sensor is determined by the ratio of the change in resonance wavelength to a small change in the refractive index or molar concentration of the analyte. The mathematical representation of this sensitivity is as follows:

$$S = \frac{\Delta\lambda_{\text{res}}}{\Delta n_a} \text{ or, } S = \frac{\Delta\lambda_{\text{res}}}{\Delta\left(\frac{V_a}{V}\right)} \tag{7.17}$$

The symbol "$\Delta\lambda_{\text{res}}$" represents the change in wavelength of the resonant sensing signal, while "Δn_a" represents the change in refractive index in a hydrated medium. "V_a" denotes the volume fraction of mono-alcohol in water, and "V" represents the volume of a pseudo-binary mixture. The parameter of detection accuracy (D.A.), which represents the true value of the measured quantity, is also highly significant. The D.A. (Detection Accuracy) of this sensor is determined by calculating the ratio between the change in wavelength of the resonant sensing signal and the Full Width at Half Maximum (FWHM) of the resonant sensing signal.

$$D.A. = \frac{\Delta\lambda_{\text{res}}}{\Delta\lambda_{0.5}} \tag{7.18}$$

The term "$\Delta\lambda_{0.5}$" refers to the spectral width at which the transmittance is at 50%. The quality parameter (Q.P.) is a significant performance metric that indicates the overall performance of the sensor being evaluated. It is determined by the essential sensitivity parameter, D.A. The definition of Q.P. is as follows:

$$Q.P. = \frac{S}{\Delta\lambda_{0.5}} \quad (7.19)$$

In Eq. (7.17), "*S*" represents sensitivity, while $\Delta\lambda_{0.5}$ refers to the full width at half maximum (FWHM) of the sensing signal. Based on the study's findings, it is recommended to have a very narrow Full Width at Half Maximum (FWHM) for the sensing signal in order to avoid inaccuracies when detecting the wavelength of the sensing signal.

7.6 Numerical Results and Discussion

The optofluidic BFAS adulteration sensor system, which includes a geometrical fault, is built and theoretically modeled. Subsequently, it is compared with existing findings in order to optimize the adulterants in the temperature-dependent molar concentration of hydrated mono-alcohols (namely methanol and ethanol) used as fuel. The current sensor configuration is designed based on the wavelength $\lambda o = 1550$ nm of the electromagnetic wave. The central structure of BFs is sufficiently empty to serve as an optofluidic flow cell and retain a pseudo-binary mixture of hydrated mono-alcohols. The circumference of the hollow core is maintained at a size that allows for easy retention of the mixture. In addition, the central part is surrounded by multiple layers of Arsenic tri-selenide, a material with a high refractive index ($n_H = 2.82$), and Poly-ether-imide PEI, a material with a low refractive index ($n_L = 1.66$) [22, 23]. The thicknesses of these layers are determined by the quarter-wave stack condition, which states that the product of the refractive index and thickness of each layer should be equal to a quarter of the wavelength $n_H d_H = n_L d_L = n_D d_D = \lambda o/4$. In addition, a defect layer made of a low-index material called Poly-ether-imide, with a refractive index of 1.66 (n_D), and a thickness of $d_D = \frac{\lambda o}{4n_d}$, is inserted in the middle of a periodic stacking consisting of "16" layers. This insertion breaks the symmetry and creates asymmetry in the arrangement of the concentric cladding layers. The glass/polymer combination was selected for this inquiry due to its convenient fabrication capabilities for the BFAS device setup. The polymer PEI exhibits excellent thermal stability, maintaining its structural integrity at temperatures as high as 400 °C [40]. Furthermore, it possesses a glass transition temperature exceeding 220 °C in the visible and near-infrared range. This high glass transition temperature is crucial for accurately measuring the temperature-dependent transmission spectra, especially within the limited temperature range of 50–500 °C considered in this investigation. The data obtained from this investigation are shown in Tables 7.3a and Table 7.3b, and visualized in Figs. 7.4 and 7.5. The results of the BFAS device configuration are presented in Table 7.4. Considering the significant influence of temperature, which can fluctuate as a result of various weather conditions and over the course of the day, Figure 7.2a–c, as well as Figure 7.3a–c were

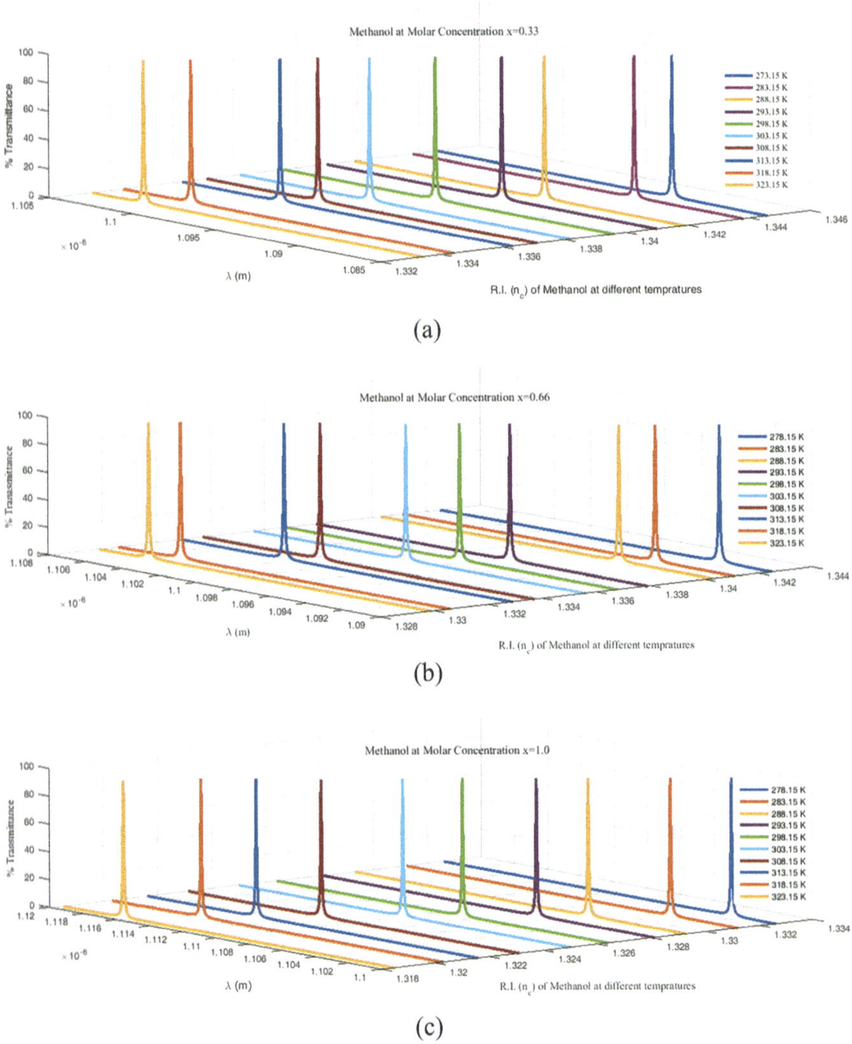

Fig. 7.2 Estimation of shift in SSP for the different molar fractions (**a**) 0.33, (**b**) 0.66 and (**c**) 1.0 of hydrated methanol fuel at different temperatures by using BFAS device setup transmittance spectra [37]. (Reprinted from Heliyon, 8, Nitesh K. Chourasia, Narendra Bihari, Ritesh Kumar Chourasia, An optofluidic Bragg fiber sensor for estimating adulterants in a temperature-dependent molar fraction of hydrated mono-alcohol fuels, e10532., Copyright 2022, with permission from Elsevier)

utilized to approximate the positions of PPBG and SSP for methanol and ethanol fuels in a pseudo-binary hydrated mixture at specific temperatures. The optical spectra shown above were generated using Eqs. 7.10 and 7.11 and referencing Tables 7.1a, 7.1b, and 7.2. The derived solid-state parameters (SSPs) for all three molar fractions ($x = 0.33$, 0.66, and 1.00) at varied temperatures (50–500 °C) have been optimized and are presented in Tables 7.3a and 7.3b. These SSPs, which have an FWHM of approximately 0.1 nm, are essential for current sensing applications.

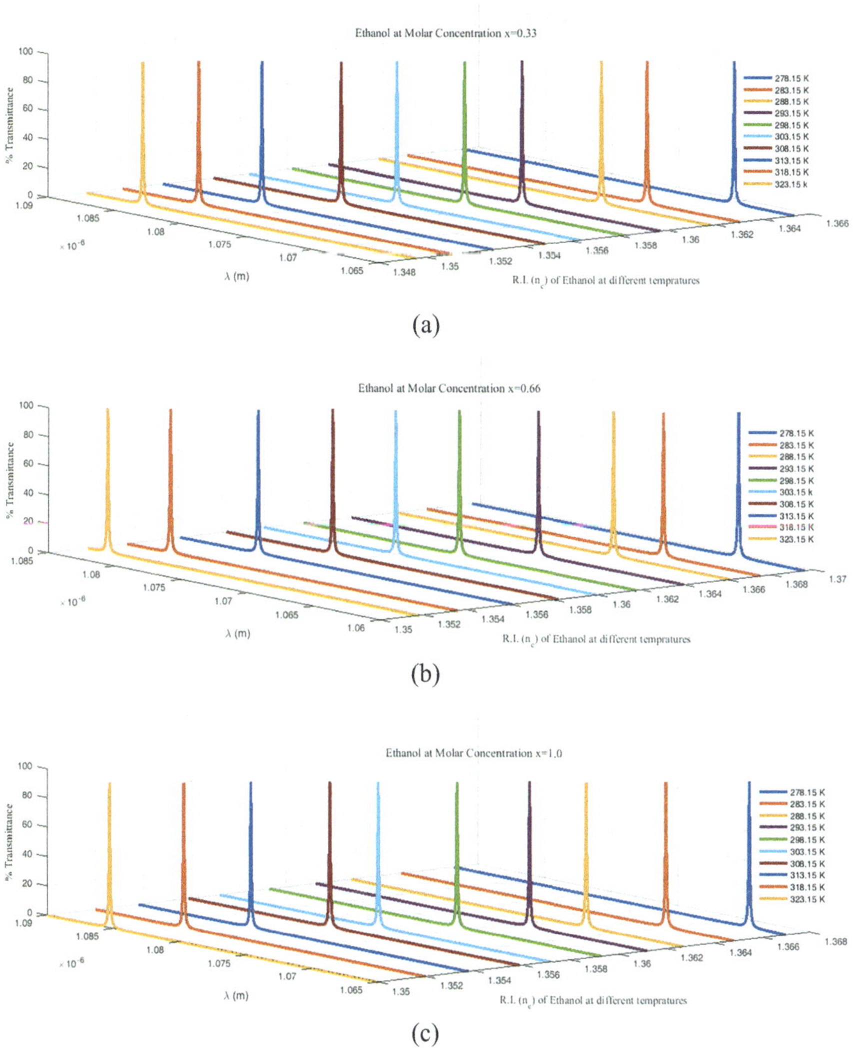

Fig. 7.3 Estimation of shift in SSP for the different molar fractions (**a**) 0.33, (**b**) 0.66 and (**c**) 1.0 of hydrated Ethanol fuel at different temperatures by using BFAS device setup transmittance spectra [37]. (Reprinted from Heliyon, 8, Nitesh K. Chourasia, Narendra Bihari, Ritesh Kumar Chourasia, An optofluidic Bragg fiber sensor for estimating adulterants in a temperature-dependent molar fraction of hydrated mono-alcohol fuels, e10532., Copyright 2022, with permission from Elsevier)

Moreover, the findings indicate that there is a noticeable redshift in the SSPs with the smallest FWHM of 0.1 nm for various molar fractions at varying temperatures. This redshift will be utilized to enhance the performance and precision of the BFAS sensor device configuration, which currently has a geometric defect. Figures 7.4 and 7.5 depict the relationship between the wavelength of SSP and different molar percentages of methanol and ethanol within the specified temperature range. This

Table 7.1a Experimental measurements data of density (ρ) and mixture RI's (n_D) for methanol (fuel) + water pseudo-binary mixtures (hydrated mono-alcohol) along with the values obtained for n_G, n_L,Δn_G, ΔR, and V^E at a temperature of 298.15 K [13]

S. no.	x	ρ (kg. m^{-3})	n_D	n_G	n_L	Δn_G	$\Delta R \times 10^{-6}$ (m^3. mol^{-1})	$V^E \times 10^{-6}$ (m^3. mol^{-1})
1	0.0000	997.02	1.3326	1.33260	1.33260	0.00000	0.00000	0.00000
2	0.3330	917.74	1.3396	1.34063	1.32931	−0.00103	−0.04050	−0.89700
3	0.6635	840.72	1.3360	1.33538	1.32737	0.00062	−0.01115	−0.84819
4	1.0000	786.62	1.3264	1.32637	1.32640	0.00000	0.00000	0.00000

Reprinted from Heliyon, 8, Nitesh K. Chourasia, Narendra Bihari, Ritesh Kumar Chourasia, An optofluidic Bragg fiber sensor for estimating adulterants in a temperature-dependent molar fraction of hydrated mono-alcohol fuels, e10532., Copyright 2022, with permission from Elsevier

Table 7.1b Experimental measurements data of density (ρ) and mixture RI's (n_D) for ethanol (fuel) + water pseudo-binary mixtures (hydrated mono-alcohol) along with the values obtained for n_G, n_L,Δn_G, ΔR, and V^E at a temperature of 298.15 K [13]

S. no.	x	ρ (kg. m^{-3})	n_D	n_G	n_L	Δn_G	$\Delta R \times 10^{-6}$ (m^3. mol^{-1})	$V^E \times 10^{-6}$ (m^3. mol^{-1})
1	0.0000	997.02	1.3326	1.33260	1.33261	0.00000	0.00000	0.00000
2	0.3332	894.81	1.3582	1.36071	1.34914	−0.00251	−0.06492	−1.02180
3	0.6664	825.77	1.3618	1.36111	1.35685	0.00069	−0.00134	−0.67725
4	1.0000	785.08	1.3593	1.35930	1.35892	0.00000	0.00000	0.00000

Reprinted from Heliyon, 8, Nitesh K. Chourasia, Narendra Bihari, Ritesh Kumar Chourasia, An optofluidic Bragg fiber sensor for estimating adulterants in a temperature-dependent molar fraction of hydrated mono-alcohol fuels, e10532., Copyright 2022, with permission from Elsevier

Table 7.2 The RI variations with the temperature of the soluble mono-alcohol fuel for each of the specified molar fractions (x)

S. no.	T (K)	Methanol (x) + Water			Ethanol(x) + Water		
		$x = 0.33$	$x = 0.66$	$x = 1.0$	$x = 0.33$	$x = 0.66$	$x = 1.0$
1.	**278.15**	1.3446	1.3425	1.3326	1.3653	1.3696	1.3670
2.	**283.15**	1.3438	1.3411	1.3312	1.3630	1.3675	1.3648
3.	**288.15**	1.3419	1.3403	1.3293	1.3618	1.3661	1.3627
4.	**293.15**	1.3410	1.3379	1.3281	1.3527	1.3640	1.3612
5.	**298.15**	1.3396	1.3368	1.3264	1.3582	1.3618	1.3593
6.	**303.15**	1.3382	1.3356	1.3250	1.3564	1.3600	1.3572
7.	**308.15**	1.3371	1.3337	1.3231	1.3549	1.3582	1.3559
8.	**313.15**	1.3363	1.3329	1.3216	1.3528	1.3561	1.3538
9.	**318.15**	1.3344	1.3306	1.3203	1.3511	1.3536	1.3520
10.	**323.15**	1.3334	1.3299	1.3185	1.3496	1.3518	1.3500

Reprinted from Heliyon, 8, Nitesh K. Chourasia, Narendra Bihari, Ritesh Kumar Chourasia, An optofluidic Bragg fiber sensor for estimating adulterants in a temperature-dependent molar fraction of hydrated mono-alcohol fuels, e10532., Copyright 2022, with permission from Elsevier

research demonstrates a direct relationship between changes in wavelength of SSP and different molar fractions at different temperatures. The linear relationship will ensure that the BFAS sensor device maintains a consistent sensitivity, which is crucial for optimal sensor performance. This information is presented in Table 7.4, and a comparison of the performance of different mono-alcohols is illustrated in Fig. 7.6.

Table 7.3a SSPs measurement for the various molar fraction of Methanol at different temperatures

		Methanol ($x = 0.33$)		Methanol ($x = 0.66$)		Methanol ($x = 1.0$)	
S. no.	T (K)	R.I.	SSP (nm)	R.I.	SSP (nm)	R.I.	SSP (nm)
1	**278.15**	1.3446	1092.00	1.3425	1093.01	1.3326	1103.01
2	**283.15**	1.3438	1092.04	1.3411	1094.04	1.3312	1104.04
3	**288.15**	1.3419	1093.08	1.3403	1095.01	1.3293	1106.00
4	**293.15**	1.3410	1094.05	1.3379	1097.09	1.3281	1107.05
5	**298.15**	1.3396	1096.00	1.3368	1099.08	1.3264	1109.01
6	**303.15**	1.3382	1097.04	1.3356	1100.02	1.3250	1110.04
7	**308.15**	1.3371	1098.05	1.3337	1102.02	1.3231	1112.02
8	**313.15**	1.3363	1099.03	1.3329	1103.00	1.3216	1113.04
9	**318.15**	1.3344	1101.09	1.3306	1104.03	1.3203	1115.02
10	**323.15**	1.3334	1103.00	1.3299	1105.01	1.3185	1116.04

Table 7.3b SSP measurement for the various molar fraction of Ethanol at different temperatures

		Ethanol ($x = 0.33$)		Ethanol ($x = 0.66$)		Ethanol ($x = 1.0$)	
S. No.	T (K)	R.I.	SSP (nm)	R.I.	SSP (nm)	R.I.	SSP (nm)
1	**278.15**	1.3653	1070.01	1.3696	1065.06	1.3670	1068.01
2	**283.15**	1.3630	1072.02	1.3675	1067.04	1.3648	1070.04
3	**288.15**	1.3618	1073.01	1.3661	1069.00	1.3627	1072.01
4	**293.15**	1.3597	1075.03	1.3640	1071.01	1.3612	1074.05
5	**298.15**	1.3582	1077.05	1.3618	1073.03	1.3593	1076.03
6	**303.15**	1.3564	1079.04	1.3600	1075.00	1.3572	1078.04
7	**308.15**	1.3549	1081.01	1.3582	1077.01	1.3559	1079.05
8	**313.15**	1.3528	1082.09	1.3561	1079.02	1.3538	1081.02
9	**318.15**	1.3511	1084.06	1.3536	1082.08	1.3520	1083.01
10	**323.15**	1.3496	1086.07	1.3518	1084.01	1.3500	1085.04

Based on the data shown in Table 7.4 and Fig. 7.6, it is evident that the highest sensitivity seen is 1057.3 nm/*RIU* when using ethanol fuel, as opposed to methanal fuel. Furthermore, we performed a direct comparison between our acquired data and the fixed temperature molar concentration measurement of impurities in hydrated mono-alcohol fuel employing a surface plasmon resonance optic sensor, as previously reported by Kumar et al. [21]. The sensitivity and practical feasibility of our acquired result in implementing the surface plasmon sensor surpasses their obtained result. The optofluidic BFAS device setup can be utilized to optimize the presence of impurities in hydrated mono-alcohols, which is dependent on temperature. In addition, the other sensing power processors, such as Q.P. and D.A., exhibit a strong inverse relationship with the full width at half maximum (FWHM) of the SSP. This

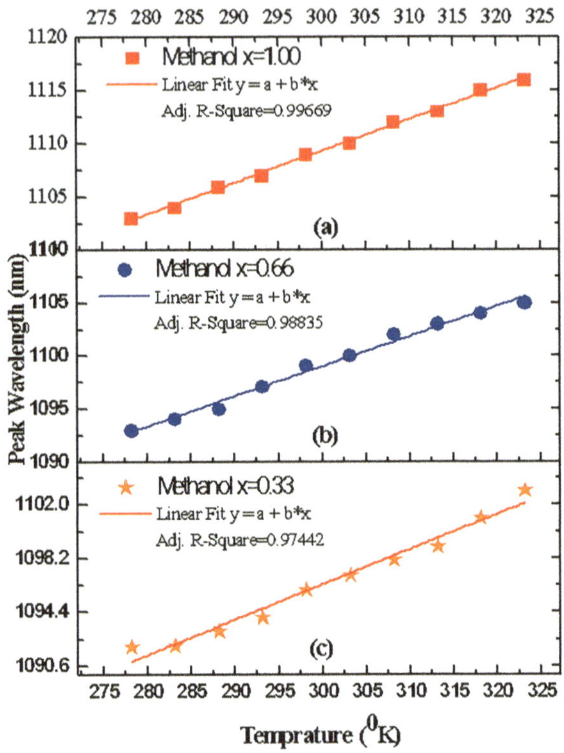

Fig. 7.4 Signature of SSP wavelength vs. temperature at a different molar concentration of methanol in pseudo-binary hydrated fuel. (Reprinted from Heliyon, 8, Nitesh K. Chourasia, Narendra Bihari, Ritesh Kumar Chourasia, An optofluidic Bragg fiber sensor for estimating adulterants in a temperature-dependent molar fraction of hydrated mono-alcohol fuels, e10532., Copyright 2022, with permission from Elsevier)

is demonstrated by Eqs. 7.18 and 7.19. These processors yield superior results compared to the findings in reference [21] due to their ability to achieve the smallest FWHM value of the SSP. As far as we know, we are the first to introduce this particular optofluidic adulteration sensor configuration with improved performance parameters. Moreover, our setup is more user-friendly and easily accessible compared to existing alternatives.

7.7 Future Scope and Experimental Realizations

From an experimental realization and future scope perspective, one research introduces a fiber optic sensor based on Lossy Mode Resonance (LMR) [41] for detecting biofuel adulteration. Here, the sensing layer was made of Tin Oxide (SnO_2) that had been deposited onto the clad-removed section of the optical fiber by means of the Radio Frequency Magnetron sputtering process. Different combinations of the analytes, such as methanol in ethanol, ethanol in methanol, methanol in water, ethanol in water, water in methanol, and water in ethanol, were tested using the manufactured sensor probe. Two LMR peaks, LMR1 and LMR2, were seen in the

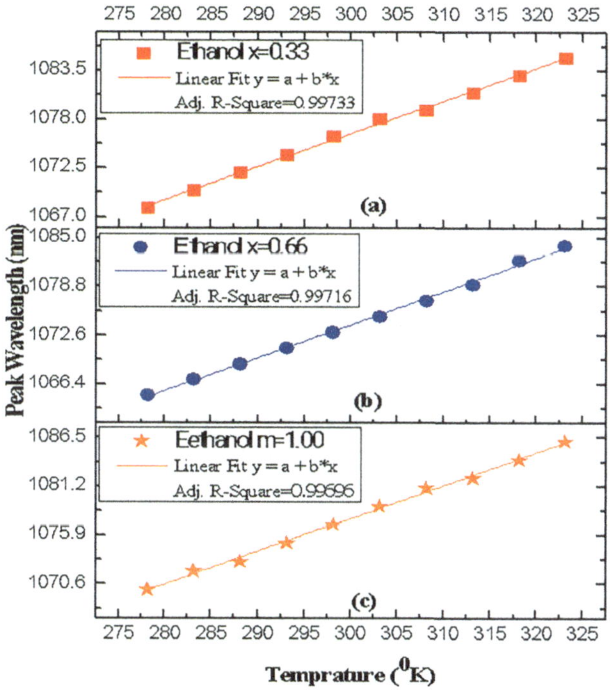

Fig. 7.5 Signature of SSP wavelength vs. temperature at a different molar concentration of ethanol in pseudo-binary hydrated fuel. (Reprinted from Heliyon, 8, Nitesh K. Chourasia, Narendra Bihari, Ritesh Kumar Chourasia, An optofluidic Bragg fiber sensor for estimating adulterants in a temperature-dependent molar fraction of hydrated mono-alcohol fuels, e10532., Copyright 2022, with permission from Elsevier)

Table 7.4 Sensing P.P.s of optofluidic BFAS device setup for optimizing the temperature-dependent molar concentration of mono-alcohols in hydrated pseudo-binary fuel mixtures

Sensing parameters	Methanol			Ethanol		
	$x = 0.33$	$x = 0.66$	$x = 1.00$	$x = 0.33$	$x = 0.66$	$x = 1.00$
(S) (nm/RIU)	982.14	952.38	943.26	1057.32	1039.32	1017.65
(D.A.)	110.0	120.0	133.0	166.0	185.0	173.0
Q.P. (RIU^{-1})	9.82×10^3	9.52×10^3	9.43×10^3	1.05×10^4	1.03×10^4	1.01×10^4

Reprinted from Heliyon, 8, Nitesh K. Chourasia, Narendra Bihari, Ritesh Kumar Chourasia, An optofluidic Bragg fiber sensor for estimating adulterants in a temperature-dependent molar fraction of hydrated mono-alcohol fuels, e10532., Copyright 2022, with permission from Elsevier

absorption spectra of the manufactured sensor probe. These peaks showed wavelength shifts when the analyte concentrations changed. For each of the analyte combinations listed above, the maximum sensor responses were 4.95%, 3.30%, 3.00%, 3.56%, 2.91%, and 3.57%, respectively. This sensor probe has sensitivity values of 0.46 nm/vol%, 0.17 nm/vol%, 0.49 nm/vol%, 0.58 nm/vol%, 0.08 nm/vol %, and 0.4 nm/vol% for these analytes, in that order. Using sensors covered with metal oxides, the real-time samples were evaluated. Findings show that probes coated with SnO_2 are more sensitive than uncoated ones.

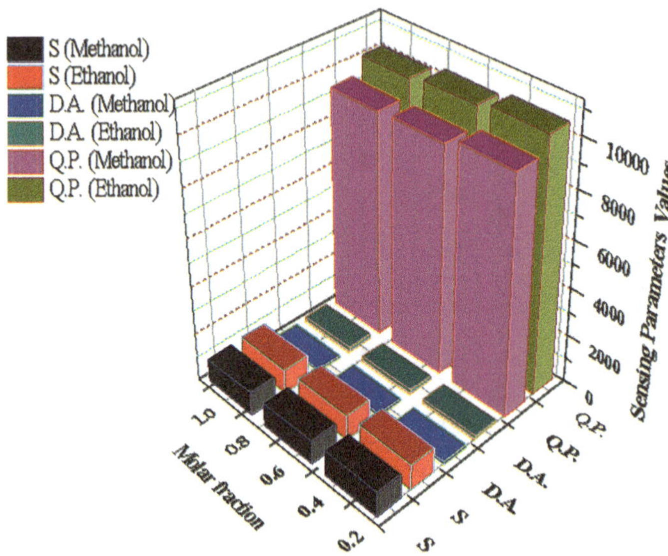

Fig. 7.6 Representation of sensing P.P's metrics of proposed optofluidic BFAS setup. (Reprinted from Heliyon, 8, Nitesh K. Chourasia, Narendra Bihari, Ritesh Kumar Chourasia, An optofluidic Bragg fiber sensor for estimating adulterants in a temperature-dependent molar fraction of hydrated mono-alcohol fuels, e10532., Copyright 2022, with permission from Elsevier)

On a global scale, edible oil is vital to dietary practices. It gives food its distinctive flavor while simultaneously supplying the nutrients our bodies require. Public health problems resulting from unqualified edible oil happen annually, particularly in low-income communities, and this is because the quality of edible oil has a direct impact on our health. Present edible oil detection methods, however, sometimes necessitate cumbersome apparatus and a plethora of chemicals, which are both expensive and environmentally harmful to less developed regions. This study focuses on the design and manufacturing of a disposable sensor [42] that can detect edible oils with high sensitivity. The sensor uses an arched optofluidic waveguide to measure the edible oils' refractive index (RI). The measurements of the edible oils' RI may be taken by our sensor with a sensitivity of 16.175 mW/RIU and a low-noise-equivalent detection limit (NEDL) of 4.099×10^{-5} RIU concurrently, across a dynamic range of 1.45839–1.47399 RIU, according to the experimental results. Furthermore, this small sensor can identify a sample in just as 30 s, making it an incredibly practical and effective tool for edible oil analysis in a range of environments. The sensor can differentiate between several kinds of commercial edible oil owing to its exceptional RI sensing performance. Nutritional and toxicological assessments may also benefit from the sensor's ability to precisely track the edible oil's RI change as frying time increases. A sensor with the potential to improve human health by offering a low-cost, high-sensitivity edible oil detection solution in developing nations.

Once again, no-core fiber's (NCF) [43] distinctive structural feature has expanded sensing applications based on multimode interference self-imaging, evanescent wave absorption, and materials coating. These applications are extensively utilized for measuring a wide range of physical, chemical, and biological parameters. With a focus on measurement techniques, structural enhancements, and the integration of sensitive materials, this overview of NCF-based sensing applications spreads out. There is an exhaustive description of many NCF-based sensor prototypes. This section begins with a brief overview of the many NCF-based sensors that use distinct analytical principles, followed by a comparison of the benefits of various structural improvements, such as fiber taper, lateral-offset fusing, and internal waveguide writing. In addition, we cover the topic of cascading NCF with fiber gratings or other structures to enable multi-parameter measurement and expand its application scope. After that, we cover how NCF can detect physical and chemical characteristics by combining sensitive materials or by coating metal films to trigger surface plasmon resonance. To back up our claims about the potential value of NCF-based sensors in biosensing, we provide several examples of recent research. Additionally, the sensing properties of several NCF structure applications are presented. The article provides a comprehensive overview of the sensing applications, discusses the challenges, and highlights the research direction for NCF-based sensor development.

7.8 Conclusion

This chapter presents the successful development of an optofluidic BFAS adulteration sensor device setup that includes a geometrical fault. The purpose of this setup is to optimize and monitor the adulterants in a temperature-dependent molar proportion of hydrated mono-alcohol fuel under different weather circumstances. The observed PPBG has a narrow spectral sensitivity profile (SSP) due to a geometrical error in the design of the BFAS sensor device. This defect is very responsive to changes in the refractive index (R.I.) of the fuel-filled core when it is hydrated. The new optofluidic BFAS device arrangement exhibits significantly enhanced P.P. metrics in comparison to previous comparable sensors, primarily due to the shortest full width at half maximum (FWHM) of the single scattering peak (SSP). The BFAS device configuration has achieved a peak sensitivity of 1057.3 nm/RIU under all weather conditions. The sensitivity of this sensor exceeds that of a surface plasmon resonance optic sensor in detecting hydrated mono-alcohols, but only when the temperature is constant. Reducing the full width at half maximum (FWHM) of the sensor signal profile (SSP) in this arrangement improves the detection accuracy (D.A.) and quantification precision (Q.P.), as these metrics are inversely related to the FWHM. Specially the last section has particularly reviewed and elaborated other experimental realizations of many recent similar type research works on optofluidic adulteration sensors which can be performed and realized also using Bragg fibers.

References

1. C. Sayin, Engine performance and exhaust gas emissions of methanol and ethanol–diesel blends. Fuel **89**, 3410–3415 (2010)
2. D.C. Rakopoulos, C.D. Rakopoulos, D.T. Hountalas, E.C. Kakaras, E.G. Giakoumis, R.G. Papagiannakis, Investigation of the performance and emissions of bus engine operating on butanol/diesel fuel blends. Fuel **89**(10), 2781–2790 (2010)
3. R. Erdiwansyah, M.S.M. Mamat, M.S.M. Sani, K. Sudhakar, A. Kadarohman, R.E. Sardjono, An overview of higher alcohol and biodiesel as alternative fuels in engines. Energy Rep. **5**, 467–479 (2019)
4. I.A. Omar, R. Mamat, O.M. Ali, N.A.C. Sidik, T. Yusaf, K. Kadirgama, M. Kettner, Alcohol and ether as alternative fuels in spark ignition engine: A review. Renew. Sust. Energ. Rev. **82**(3), 2586–2605 (2018)
5. M.A. Ghadikolaei, C.S. Cheung, K.F. Yung, Study of combustion, performance, and emissions of a diesel engine fueled with diesel/biodiesel/alcohol blends having the same oxygen concentration. Energy **157**, 258–269 (2018)
6. Y.N. Guragain, K.V. Probst, P.V. Vadlani, Fuel Alcohol Production, in *Encyclopedia of Food Grains*, ed. by C. Wrigley, H. Corke, K. Seetharaman, J. Faubion, 2nd edn., (Academic Press, 2016), pp. 235–244
7. IEA, *Renewables 2021* (IEA, Paris, 2021) https://www.iea.org/reports/renewables-2021
8. https://www.niti.gov.in/expert-committee-roadmap-ethanol-blending-india-2025
9. N.P. Cheremisinoff, *Industrial Solvents Handbook*, 2nd edn. (Marcel Dekker, New York, 2003)
10. M. Zhang, Y. Ge, X. Wang, D. Thomas, S. Su, H. Li, An assessment of how bio-E10 will impact the vehicle-related ozone contamination in China. Energy Rep. **6**, 572–581 (2020)
11. R.J. Riobóo, M. Philipp, M.A. Ramos, et al., Concentration and temperature dependence of the refractive index of ethanol-water mixtures: Influence of intermolecular interactions. Eur. Phys. J. E **30**, 19 (2009)
12. N.A.S. Ramli, F. Abdullah, Study of density, surface tension, and refractive index of binary mixtures containing alkyl levulinate and n-alcohol from 298.15 to 323.15 K. J. Chem. Eng. Data **66**(5), 1856–1876 (2021)
13. J.V. Herráez, R. Belda, Refractive indices, densities and excess molar volumes of monoalcohols + water. J. Solut. Chem. **35**, 1315–1328 (2006)
14. J. Ortega, Densities and refractive indices of pure alcohols as a function of temperature. J. Chem. Eng. Data **27**(3), 312–317 (1982)
15. A.C. Silva, L.F.B.L. Pontes, M.F. Pimentel, M.J.C. Pontes, Detection of adulteration in hydrated ethyl alcohol fuel using infrared spectroscopy and supervised pattern recognition methods. Talanta **93**, 129–134 (2012)
16. G.F. Giordano, D.C.M. Ferreira, T.R. de Carvalho, L.C.S. Vieira, M.H. de Oliveira Piazzetta, R.S. Lima, A.L. Gobbia, Portable platform for rapid and indirect photometric determination of water in ethanol fuel samples. Anal. Methods **6**(23), 9497–9502 (2014)
17. W.A. De Oliveira, C. Pasquini, Determination of water in ethanol and acetone by direct injection enthalpimetry based on the heat of dilution. Talanta **31**, 82 (1984)
18. C.R. Omido, S.L. Oliveira, J.R. Lima, K.F. Magalhães, A.A. de Carvalho, C. Kitano, Water content in hydrated ethanol fuel measured by a photothermal chamber with a transparent transducer. Fuel **157**(1), 122–125 (2015)
19. H.V. Dantas, M.F. Barbosa, A. Pereira, M.J.C. Pontes, P.N.T. Moreira, M.C.U. Araújo, An inexpensive NIR LED webcam photometer for detection of adulterations in hydrated ethyl alcohol fuel. Microchem. J. **135**, 148–152 (2017)
20. M.K.K. Figueiredo, R.P.B. Costa-Felix, L.E. Maggi, A.V. Alvarenga, G.A. Romeiro, Biofuel ethanol adulteration detection using an ultrasonic measurement method. Fuel **91**, 209 (2012)
21. S. Kumar, G. Sharma, V. Singh, Modeling of surface plasmon resonance sensor for detection of mass concentration of ethanol and methanol in a binary mixture. Infrared Phys. Technol. **67**, 190–196 (2014)

22. R.K. Chourasia, V. Singh, Estimation of photonic bandgap in the hollow core cylindrical multilayer structure. Superlattice. Microst. **116**, 191–199 (2018)
23. R.K. Chourasia, S. Prasad, V. Singh, Performance study of a liquid-core Bragg fiber sensor in the presence of a defect layer. Opto-Electron. Rev. **25**, 215–221 (2017)
24. N. Bihari, A. Srivastava, N.K. Chourasia, R.K. Chourasia, H-polarised EM-wave transport in polymeric-chalcogenide columnar photonic materials. J. Phys. Conf. Ser. **2426**, 012009 (2023)
25. R.K. Chourasia, N.K. Chourasia, N. Bihari, *Optical Properties of Hollow-Core Bragg Fiber Waveguides*, vol 23 (Benthem Science, 2023), pp. 214–236
26. R.K. Chourasia, V. Singh, *Comparative analysis of the dispersion characteristic of Bragg fibers having different refractive index contrast*, in 13th International Conference on Fiber Optics and Photonics, OSA Technical Digest (online) (Optica Publishing Group, 2016), paper Th3A.75
27. R.K. Chourasia et al., Exploring optical properties in cylindrical polymeric-chalcogenides photonic materials. Mater. Today Proc. **67**(5), 625–631 (2022)
28. N.K. Chourasia, A. Srivastava, V. Kumar, R.K. Chourasia, Doubly electrically tuned cylindrical Bragg fiber waveguide inline optical filter for multiwavelength LASER applications. Mater. Today Commun. **25**(101620) (2020)
29. R.K. Chourasia, C.S. Yadav, A. Upadhyay, N.K. Chourasia, V. Singh, Analysis of Bragg fiber waveguides having a defect layer for biosensing application. Optik **200**, 163400 (2020)
30. R.K. Chourasia, C.S. Yadav, A. Upadhyay, N.K. Chourasia, V. Singh, Comparative study of defect mode intensity and wavelength modulation in Bragg fiber waveguide sensors. Optik **204**, 164198 (2020)
31. R.K. Chourasia, S. Prasad, V. Singh, A novel Bragg fiber waveguide based narrow band inline optical filter. Optik **169**, 269–276 (2018)
32. N.K. Chourasia, A. Srivastava, V. Kumar, R.K. Chourasia, Comparative performance study of liquid core cylindrical Bragg fibre waveguide biosensors. Pramana J. Phys. **95**, 9 (2021)
33. R.K. Chourasia, S. Prasad, V. Singh, Voltage-tunable passband in a cylindrical multilayered structure containing PMMA and 0.67 PMN–0.33 PT single crystal as a defect layer. Opt. Quant. Electron. **48**(12), 539 (2016)
34. R.K. Chourasia, V. Singh, *Cylindrical Bragg waveguide based temperature sensors*, in IEEE Uttar Pradesh Section International Conference on Electrical, Computer and Electronics Engineering (UPCON), Varanasi, 2016, pp. 87–90. https://doi.org/10.1109/UPCON.2016.7894630
35. N.K. Chourasia, A. Srivastava, V. Kumar, R.K. Chourasia, Optimizing temperature-dependent molar volume fraction of biodiesel fuel in a pseudo-binary mixture through Bragg fiber waveguide sensor having defect layer. Fuel **293**, 120489 (2021)
36. R.K. Chourasia, N.K. Chourasia, A. Srivastava, N. Bihari, *Photonic Nanostructured Bragg Fuel Adulteration Sensor*, vol 28 (Benthem Science, 2023), pp. 237–264
37. N.K. Chourasia, N. Bihari, R.K. Chourasia, An optofluidic Bragg fiber sensor for estimating adulterants in a temperature-dependent molar fraction of hydrated mono-alcohol fuels. Heliyon **8**(9), e10532 (2022)
38. M. Born, E. Wolf, *Principles of Optics* (Cambridge University Press, Cambridge, London, 1999)
39. A.W. Snyder, J. Love, *Optical Waveguide Theory*, 2nd edn. (Springer, 2008)
40. R. Mercado, Y. Wang, T. Flaim, W. Di Menna, U. Senapati, *Thin-film polyetherimides with controlled refractive indices*, in Proceedings of SPIE: Organic Photonic Materials and Devices VI, 5351 (2004)
41. A. Prasanth, V.K. Harini, P. Manivannan, M. Velumani, S. Narasimman, S.R. Meher, Z.C. Alex, Detection of biofuel adulterants using an optical fiber-based refractive index sensor. Optik **291**, 171345 (2023)
42. R. Yan et al., A cost-effective and high-sensitivity optofluidic refractive index sensor for rapid edible oil analysis. IEEE Sensors J. **23**(20), 24188–24195 (2023)
43. Y. Zhao, J. Zhao, Q. Zhao, Review of no-core optical fiber sensor and applications. Sensors Actuators A Phys. **313**(112160) (2020)

Chapter 8
Bragg Fiber Optoelectronic Applications: Optical Inline Filters for Multiwavelength Applications

Abstract Previous chapters have extensively discussed the various potential applications of Bragg fiber (BF) structures as sensors for detecting bio/fuel adulteration. These sensors rely on wavelength and intensity modulation in defect signals to achieve optimal sensing performance. Compared to other sensing methods such as planner photonic crystals, photonic crystal fibers, and surface plasmon sensors, BFs offer superior sensing behavior. Furthermore, we have noticed that the BF sensor has superior sensing capabilities compared to other existing methodologies or systems. In this current chapter of the book, we have examined a possible use in the field of optoelectronics. This chapter presents a detailed analysis of doubly tuned Bragg fibers that are used as low-power optical inline filters (OIFs). The filters have been modeled, designed, and studied using a voltage-tunable single-crystal PMMA polymer defect cavity. The analysis is conducted using Hankel formalism in a cylindrical coordinate system, specifically for multiwavelength LASER applications. The structure of the BFs exhibits a photonic band gap (PBG) that produces a resonant transmission signal inside the PBG area. The thickness of the PMMA fault cavity is regulated by the application of an external direct current (DC) voltage. The suggested OIFs can achieve a tunable PBG of up to 828 nm by changing the incidence angle up to 80^0. Therefore, the transmission signal that resonates is moved towards the blue end of the spectrum by 243 nm. This shift may be further adjusted by applying DC voltages, a phenomenon known as double electrically tuning. In addition, when operating in the doubly electrically tuned mode, this optoelectronic device has a significantly wide application range of up to 129 nm. This wide range may be achieved with a relatively modest change in voltage, typically about 20 V. Therefore, the current technology may be employed as a multiwavelength LASER source.

8.1 Introduction

There is a significant demand for multiwavelength LASER sources in various areas, including optical instrument testing, optical signal processing, optical communications, modern telecommunications, fiber sensor systems, and sensor multiplexing.

R. K. Chourasia, A. Katti, *Bragg Fibers*,
https://doi.org/10.1007/978-3-031-65164-9_8

These laser sources possess substantial potential in the testing of optical instruments and are also utilized in the measurement and evaluation of wavelength division multiplexing (WDM) and optical d-multiplexers (O-DMUX) components [1, 2]. The optical sources must fulfill various criteria, such as having a larger number of components/channels that cover a wide range of wavelengths, exhibiting inline behavior, providing moderate output power with a high signal-to-noise ratio, and having a narrower channel width, specifically the smallest full width at half maxima (FWHM) of the signal. These parameters are essential for achieving single-mode functioning for each laser wavelength. Accomplishing all of these needs simultaneously is a formidable task. Various scientific methodologies have been utilized to attain fiber lasers with a narrow spectral width. Researchers have utilized the liquid crystal etalon filter as an electrically adjustable wavelength tuning element in the development of a tunable fiber laser [1]. This infrared wavelength-tunable laser offers a broad variety of applications. However, it has the smallest wavelength spread and is restricted to the infrared spectrum. Hernandez et al. [2] introduced a laser that can emit many wavelengths, which is made possible by incorporating erbium-doped material into a tunable device called a Mach-Zehnder interferometer. The laser utilizes photonic crystal fiber to enhance its performance. The LASER has enhanced stability even under normal environmental conditions. However, its emission capability is restricted to generating just a single, double, triple, or quadruple line, which can be tuned within the range of 1530 nm to 1556 nm by manipulating the polarization state. In addition, numerous scientists and researchers have introduced multiwavelength LASER models that incorporate different types of filters, such as, acoustic optical filters (AOFs), fiber Bragg gratings (FBGs), Sagnac interferometers (SIs), Fabry-Pérot interferometers (FPIs), Mach-Zender interferometers (MZIs), [3–12] and so on. Multiwavelength lasers with optical filters provide distinct advantages in many fields. However, these valuable approaches are constrained by factors like size, single-mode operation, wavelength range, power consumption, inline behavior, switchability, and repeatability. Hence, the BFs structure [13–23] exhibits a promising potential and will be very advantageous in multiwavelength LASER applications. The BFs structure is built upon the periodic arrangement of concentric cylinders with varying R.Is. This configuration allows for the propagation of EM waves into the hollow core via repeated Bragg reflections from the periodic claddings [11–14]. Additionally, the structure of BFs must exhibit PBG. PBG refers to a certain range of wavelengths that are not allowed to travel through the structure of the BFs. Bragg fibers offer a wide range of potential uses in several fields such as biosensors [14, 16, 17], high-temperature sensors [15], and intensity and wavelength modulation [16]. In this chapter, the high refractive index (R.I.) contrast chalcogenide glass, Arsenic tri-selenide As_2Se_3, and polymer film are discussed. This study has focused on the use of a cylindrical multi-clad structure made of poly-ether-imide (PEI) for the fabrication of hollow-core optical fibers. In order to improve the overall performance of the suggested prospective application, it is highly preferable to have a small transmission bandwidth (specifically, a resonant transmission signal) and a consistent intensity for inline application as the output signal. In the current periodic Bragg-Fresnel structure, the desired characteristic may be obtained by including a

PMMA polymer defect cavity into the cladding periodicity. By means of this cavity, a resonant transmission signal is generated within the photonic bandgap region. By introducing a PMMA defect layer into the periodic arrangement, the periodic symmetry is disrupted, thereby producing this defect cavity. The mathematical Hankel formalism and the transfer matrix method (TMM) were utilized in order to assess the electromagnetic (EM) wave's transmission characteristics via the BFs structure. By modulating the angle of incidence, the present architectural design of BFs is capable of modifying both the photonic bandgap (PBG) and the output signal. Furthermore, the electromechanical behavior and piezoelectric properties of the ferroelectric single crystal $(1-x)\mathrm{Pb}\left(\mathrm{Mg}_{\frac{1}{3}}\mathrm{Nb}_{\frac{2}{3}}\right)O_3(PMN-xPT)$ (PMN-PT) will be utilized to accomplish the tuning procedure [24]. Multiple research works have exhaustively studied the piezoelectric and electromechanical properties and traits of these materials [25, 26].

The ferroelectric PMN-PT single crystal is of great interest for studying reproducibility and switching issues due to its fast response time, reversible domain switching mechanism, accurate control, and small device size [24, 25]. This device has the ability to produce a substantial recoverable deformation under both positive and negative direct current (DC) voltage, making it extremely beneficial for scientific investigations. Research has shown that the electromechanical coupling coefficient (k_{33}) of single crystal PMN-PT is 94%, and the piezoelectric constant (d_{33}) is 2800 pC/N for both positive and negative DC bias voltages. These values are adequate and have the ability to adjust the material thickness by roughly 2.8 nm per volt [24, 25]. Hence, this study focuses on the discussion of a cylindrical Bragg fiber structure with dual electrical tuning capabilities for the use of multiwavelength lasers. The cylindrical BFs construction is used to ensure a dependable and uninterrupted transmission signal in optical networks.

8.2 Comprehensive Device Setup of Bragg Optical Inline Filter

Figure 8.1a displays the thickness and RI profile of the hollow-core high-contrast BFs construction with a defect cavity. The suggested structure of BFs exhibits symmetry concerning the longitudinal axis, also known as the propagation axis. The current configuration consists of a repeated pattern of periodic bilayer materials. The high R.I. chalcogenide material used is Arsenic tri-Selenide (As_2Se_3), (n_H, d_H), while the low R.I. layer is made of poly-ether-imide (PEI) (n_L, d_L). In addition, a defect cavity made of a low R.I. polymer called PMMA is created by disrupting the regular pattern of layers. This cavity is described by its R.I. (n_D) and its thickness (d_D). The selection of all thicknesses in this context is based on the quarter wave stack condition, which ensures maximum transmission intensity and inline behavior. Specifically, the thicknesses of the high-index layer (n_H), low-index layer (n_L), and dielectric layer (n_D) are selected so that $n_H d_H = n_L d_L = n_D d_D = \lambda_c/4$, where λ_c

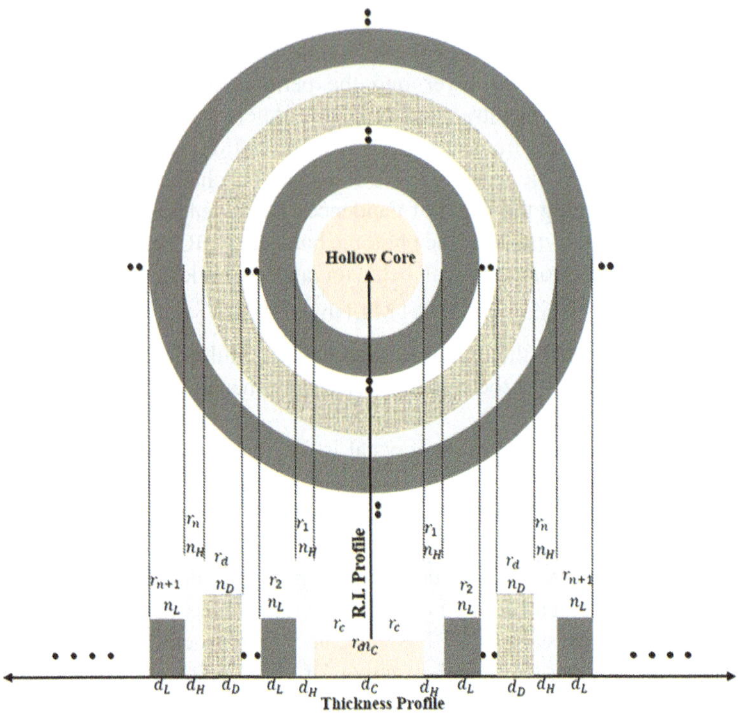

Fig. 8.1a R.I. and thickness profile of hollow-core high-contrast BFs structure with a PMMA defect cavity. (Reprinted from [11], Copyright 2020, with permission from Elsevier)

represents the design wavelength of the BFs structure. Figure 8.1b depicts a diagram showing the structure of electrically adjustable hollow-core high-contrast BFs. These BFs have a PMMA cavity and utilize a single crystal ferromagnetic material called PMN-PT on both sides. This material can change its thickness in response to an electric field, which in turn affects the resonance condition in the defect cavity. A compliant electrode, such as carbon conductive grease, is used to insert the opto-electronic piezoelectric material made of single crystal PMN-PT. This allows for controlled expansion and contraction of the PMN-PT crystal. Here, an electrode with carbon conductive grease is utilized for electrical conduction. Additionally, the current structure of the BFs is determined by the application of positive and negative DC voltage within the cylindrical symmetry on the electrode. Ultimately, the thorough examination of the BFs structure is very dependable and not subject to doubt. The Abouraddy et al. group has successfully constructed a standard version of a BFs structure with many functions, demonstrating the practicality of a voltage-tunable structure. Additionally, Fig. 8.1c depicts the diagrammatic portrayal of the implementation of the low power consumption doubly electrically adjusted hollow-core high-contrast BFs structure optical inline filter for multiwavelength LASER applications. The EM wave originating from the wider source is directed toward the central part of the BFs structure. Due to the presence of a cavity defect inside the

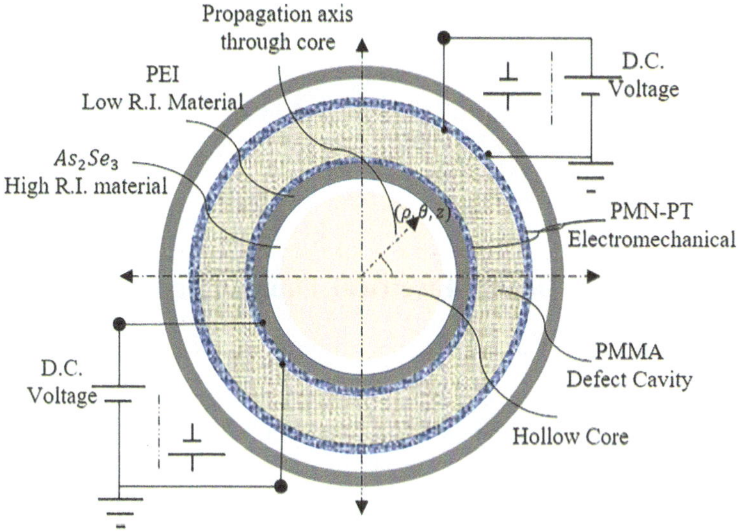

Fig. 8.1b Schematic cross-sectional view of electrically tuned hollow-core high-contrast BFs with PMMA cavity and electro-optic single crystal material PMN-PT onto its both sides [11]. (Reprinted from Materials Today Communications, 25, Nitesh K. Chourasia, Ankita Srivastava, Vinay Kumar, Ritesh Kumar Chourasia, Doubly electrically tuned cylindrical Bragg fiber waveguide inline optical filter for multiwavelength LASER applications, 101620, Copyright 2020, with permission from Elsevier)

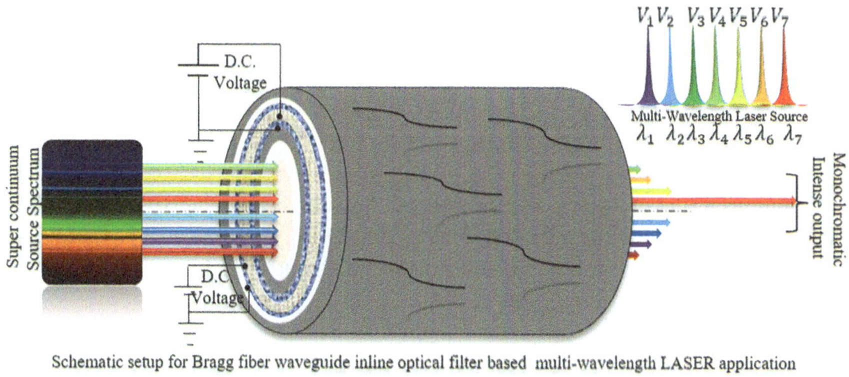

Fig. 8.1c Schematic representation of electrically tuned BFs-based optical inline filter for multiwavelength LASER applications [11]. (Reprinted from Materials Today Communications, 25, Nitesh K. Chourasia, Ankita Srivastava, Vinay Kumar, Ritesh Kumar Chourasia, Doubly electrically tuned cylindrical Bragg fiber waveguide inline optical filter for multiwavelength LASER applications, 101620, Copyright 2020, with permission from Elsevier)

intended structure, a highly focused and intense transmission peak has emerged within the wavelength range of a PBG in the transmission spectra. This peak may be largely adjusted by modifying the angle of incidence, resulting in the generation of many intense wavelengths. Additionally, the resonance state of the cavity may be modified by applying direct current (DC) voltages by electrical methods, allowing for further precise adjustments. The current output of the BFs structure yields highly concentrated sources of many wavelengths, as seen in the picture, at various DC voltages. These sources are utilized for applications involving multiwavelength LASERs applications.

8.3 Modeling of Doubly Electrical Tunable Bragg Optical Inline Filter

The comprehensive explanation of the proposed BFs structure has already been addressed in Sect. 8.2. In addition, the current Sect. 8.3 provides a detailed explanation of the theoretical modeling of the BFs structure. This is achieved by the utilization of the Transfer Matrix Method (TMM) and Hankel formalism, specifically in cylindrical coordinates. The recommended index profile for this framework is as follows:

$$\begin{aligned} n(r) = n_c,\ & 0 \le r < r_c; \\ n_H, & r_c \le r < r_1; \\ n_L, & r_1 \le r < r_2; \\ n_H, & r_{n-2} \le r < r_{n-1}; \\ n_L, & r_{n-1} \le r < r_D; \\ n_D,\ & r_D \le r < r_{D'}; \\ n_H, & r_{D'} \le r < r_{n+1}; \\ n_L, & r_{n+1} \le r < r_{n+2}; \end{aligned} \tag{8.1}$$

Solutions to the problems of reflection and transmission in the BFs structure can be accomplished using the cylindrical wave TMM [27]. The Maxwell equations are denoted as the following [28–33] for a substrate characterized by permittivity "ε" and permeability "μ."

$$\nabla \times E = -j\omega\mu H \tag{8.2}$$

$$\nabla \times H = j\omega\varepsilon E \tag{8.3}$$

The relationship between the longitudinal and transverse components of the electric and magnetic fields in the cylindrical coordinate system (r, θ, z) is defined assuming the temporal aspect of all fields as exp.($j\omega t$) [28–33]. The Helmholtz wave equations in the cylindrical coordinate system are expressed as:

$$r\frac{\partial}{\partial r}\left(r\frac{\partial H_z}{\partial r}\right) - r^2\frac{1}{\varepsilon}\frac{\partial \varepsilon}{\partial r}\frac{\partial H_z}{\partial r} + \frac{\partial}{\partial \theta}\left(\frac{\partial H_z}{\partial \theta}\right) + \omega^2\mu\varepsilon r^2 H_z = 0 \text{ (for TE – polarization)} \tag{8.4}$$

$$r\frac{\partial}{\partial r}\left(r\frac{\partial E_z}{\partial r}\right) - r^2\frac{1}{\mu}\frac{\partial \mu}{\partial r}\frac{\partial E_z}{\partial r} + \frac{\partial}{\partial \theta}\left(\frac{\partial E_z}{\partial \theta}\right) + \omega^2\mu\varepsilon r^2 E_z = 0 \text{ (for TM – polarization)} \tag{8.5}$$

By applying the variable separation method, it is possible to solve the aforementioned equations. For the radial and angular components of the cylindrical wave, the Maxwell field solution can be derived [27, 29]. Three components of the solution for TE-polarized waves (H_Z, E_θ, E_r) are nonzero, and are expressed as,

$$H_Z(r,\theta) = [AJ_m(kr) + BY_m(kr)]e^{jm\theta} \tag{8.6a}$$

$$E_\theta = \frac{1}{-j\omega\varepsilon}\frac{\partial H_z(r,\theta)}{\partial r} \tag{8.6b}$$

$$E_r = \frac{m}{\omega\varepsilon}\frac{H_z(r,\theta)}{r} \tag{8.6c}$$

where we let A and B represent arbitrary constants. J_m and Y_m denote Bessel and Neumann functions respectively. The wave vector of the medium is represented by $k = \left(\frac{\omega}{c}\right) n$, where ω is the angular frequency and c is the velocity of light in free space. The refractive index of the medium is denoted by "n." The transfer matrix $\widehat{T}$ elements, which establish the correlation between the field vector and a different point r from the initial position "r_c," are determined using Abeles' multilayer BFs structure theory [27, 29]. These elements are denoted by,

$$\widehat{T} = [T_{11}\ T_{12}\ T_{21}\ T_{22}] \tag{8.7}$$

$$T_{11} = \frac{\pi}{2}kr_c[Y'_m(kr_c)J_m(kr) - J'_m(kr_c)Y_m(kr)] \tag{8.8a}$$

$$T_{21} = j\frac{\pi}{2}kr_c p[Y'_m(kr_c)J'_m(kr) - J'_m(kr_c)Y'_m(kr)] \tag{8.8b}$$

$$T_{22} = \frac{\pi}{2}kr_c[J_m(kr_c)Y'_m(kr) - Y_m(kr_c)J'_m(kr)] \tag{8.8c}$$

$$T_{12} = -j\frac{\pi}{2}\frac{kr_c}{p}[J_m(kr_c)Y_m(kr) - Y_m(kr_c)J_m(kr)] \tag{8.8d}$$

where $p = \sqrt{\frac{\mu}{\varepsilon}}$ is the medium intrinsic impedance. The steps above can then be used to write the defect transfer matrix $\widehat{T}_D$ as well. Since the element of the transfer matrix is based on the radii of two surfaces, replacing the nth layer gives us the matrix $\widehat{T}_n$. We can also use $n(r) = n_L$ if n is even and $n(r) = n_H$ if n is odd because the structure is periodic. We have a total 2 N layers in the BFs structure; hence the total transfer matrix becomes,

$$[V(r_f)\ U(r_f)\ ] = T_{2N}\ldots T_2T_1[V(r_c)\ U(r_c)\ ] = \widehat{T}[V(r_c)\ U(r_c)\ ] \tag{8.9}$$

where Helmholtz equation solution at position r is $U(r)$ and $V(r)$.

The above transfer matrices, which are the same as Fresnel's equations [27] in a flat geometry, are used to figure out how waves reflect and move through BFs structures with several layers. The coefficient of reflection for a multilayer BF can be found by

$$r_M = \frac{\left(T'_{21} - jp_0c_{m0}^{(2)}T'_{11}\right) + jp_fc_{mf}^{(2)}\left(T'_{22} - jp_0c_{m0}^{(2)}T'_{12}\right)}{\left(-T'_{21} + jp_0c_{m0}^{(1)}T'_{11}\right) + jp_fc_{mf}^{(2)}\left(-T'_{22} + jp_0c_{m0}^{(1)}T'_{12}\right)} \tag{8.10}$$

In the above term, the matrix elements T'_{11},T'_{12},T'_{21}, and T'_{22} are the inverse matrix element of matrix in Eq. (8.9) and

$$C_{ml}^{(1,2)} = \frac{H_m^{(1,2)\prime}(k_lr_l)}{H_m^{(1,2)}(k_lr_l)} \text{ with } l = 0, f \tag{8.11}$$

The first and second kind Hankel functions can be denoted as $H_m^{(1)}$ and $H_m^{(2)}$, respectively, where m represents the azimuthal field variation. At this point, reflectance is quantified as $R = |r_M|^2$, and the corresponding percentage transmittance for non-absorbing media is calculated as $T = (1 - R) \times 100$ encompassing values from 0% to 100%. The formula for the TM wave can be obtained simply by inserting $\mu \leftrightarrow \varepsilon$, $j \leftrightarrow -j$ in the TE wave expression.

8.4 Numerical Findings and Discussion

A theoretical model is developed for the BFs structure, which has a length of 15 cm and is devoid of any bending effects, using the design wavelength $\lambda_c = 1550$ nm. The center air core has an R.I. of $n_c = 1.0$ and a diameter of $d = 2r_c = 330$ μm. For transmission to be maximum, the thickness of the periodic layer As_2Se_3, the cavity layer PMMA, and the quarter wave stack requirement $n_Hd_H = n_Ld_L = n_Dd_D = \lambda_c/4$ are regulated. The respective R.I. values for As_2Se_3, PEI, and PMMA are 2.82, 1.66 [11, 14], and 1.498 [14].

$d_{PMN} = \alpha\lambda_c/2n_{PMN}$ represents the flexible thickness of the single crystal PMN-PT. The RI (n_{PMN}) of PMN-PT is 2.59 [24]. The variable "α" is associated with the piezoelectric crystal strain of PMN-PT, which is contingent upon the external electric field [26]. The piezoelectric strain constant d_{33} for ferroelectric PMN-PT is 2800 pC/N [24].

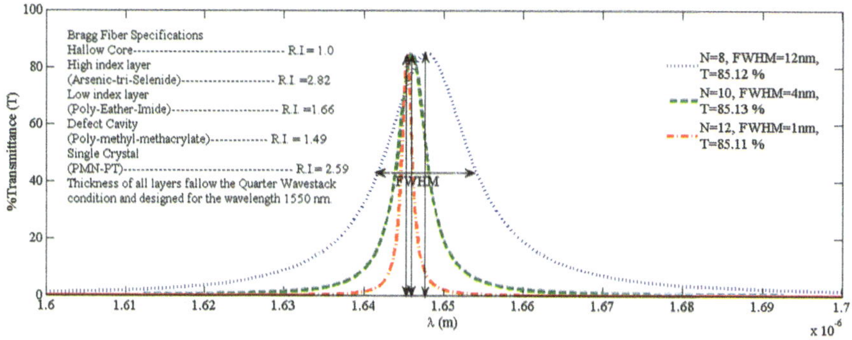

Fig. 8.2 FWHM dependency of filter signal on the no. of periodicity of BFs-based optical inline filter for multiwavelength Laser applications [11]. (Reprinted from Materials Today Communications, 25, Nitesh K. Chourasia, Ankita Srivastava, Vinay Kumar, Ritesh Kumar Chourasia, Doubly electrically tuned cylindrical Bragg fiber waveguide inline optical filter for multiwavelength LASER applications, 101620, Copyright 2020, with permission from Elsevier)

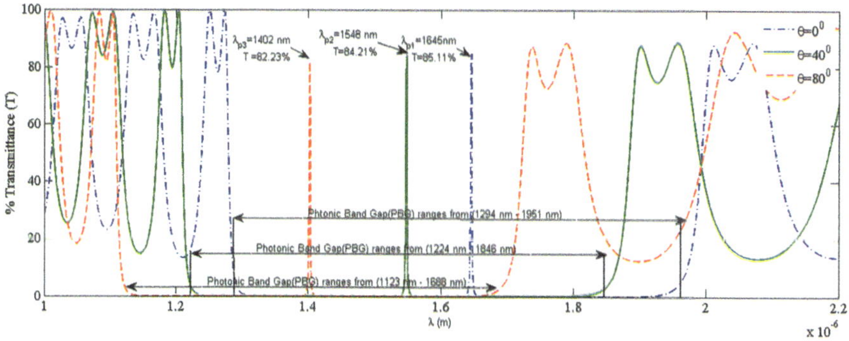

Fig. 8.3 Primary tuning in hollow-core high-contrast BFs optical inline filter PBG and filter signal by changing in angle of incidence [11]. (Reprinted from Materials Today Communications, 25, Nitesh K. Chourasia, Ankita Srivastava, Vinay Kumar, Ritesh Kumar Chourasia, Doubly electrically tuned cylindrical Bragg fiber waveguide inline optical filter for multiwavelength LASER applications, 101620, Copyright 2020, with permission from Elsevier)

The results of the simulation are presented in Figs. 8.2, 8.3, 8.4, and 8.5. These figures demonstrate the doubly electrically adjusted behavior of the hollow-core high-contrast BFs optical inline filter for LASER applications. The primary distinction between the two tuning processes is in their nature: one is mechanical, resulting from changes in the angle of incidence, while the other is electrical, arising from the modulation of a defective cavity caused by an electrical direct current (DC) voltage. The initial tuning mechanism is responsible for determining the required intervals between filter wavelengths. Once the range is determined, further refinement of the wavelength can be performed by the use of a low-power DC voltage mechanism within that specific range. Figure 8.2 demonstrates that the transmittance at the output of the BFs remains consistently high and does not vary significantly for different periodicities. Furthermore, it has been shown that the increase in the number of periodicities leads to a decrease in the flatness of the transmittance signal

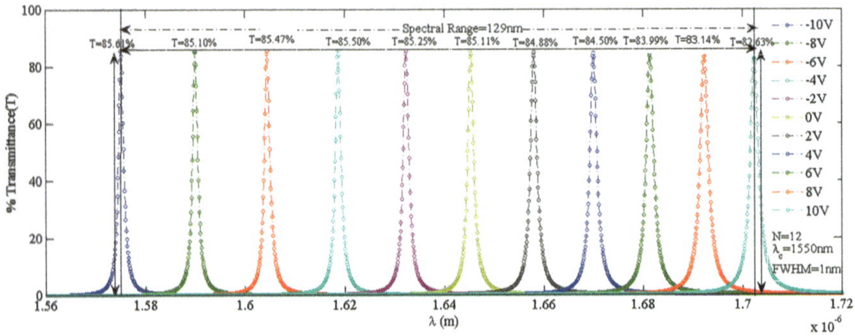

Fig. 8.4 Doubly electrical tuning in hollow-core high-contrast BFs optical inline filter signal in particular PBG span with positive and negative DC bias voltage applied on single crystal for multiwavelength LASER application [11]. (Reprinted from Materials Today Communications, 25, Nitesh K. Chourasia, Ankita Srivastava, Vinay Kumar, Ritesh Kumar Chourasia, Doubly electrically tuned cylindrical Bragg fiber waveguide inline optical filter for multiwavelength LASER applications, 101620, Copyright 2020, with permission from Elsevier)

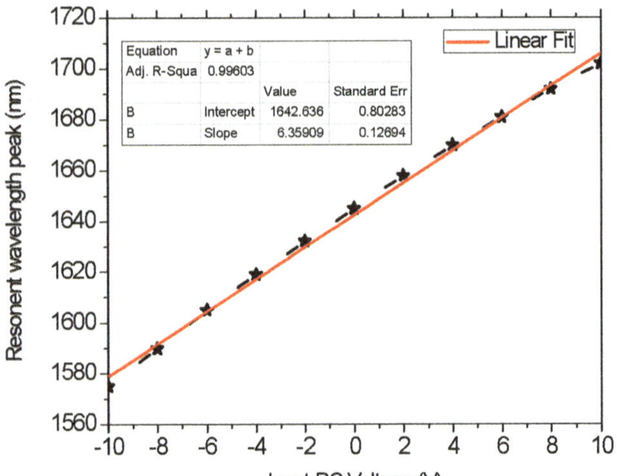

Fig. 8.5 Linear performance of BFs-based optical inline filter for multiwavelength LASER application [11]. (Reprinted from Materials Today Communications, 25, Nitesh K. Chourasia, Ankita Srivastava, Vinay Kumar, Ritesh Kumar Chourasia, Doubly electrically tuned cylindrical Bragg fiber waveguide inline optical filter for multiwavelength LASER applications, 101620, Copyright 2020, with permission from Elsevier)

detected at the output. The flatness of a curve is quantified by its Full Width at Half Maximum (FWHM). In the transition from N = 8 to N = 12 periodic layers, the FWHM reduces from 14 nm to 1 nm. The results demonstrate a significant correlation between the Full Width at Half Maximum (FWHM) of the transmittance signal and the number of periodic layers. Moreover, by adjusting the angle at which electromagnetic waves interact with the fiber core, it is possible to modify the photonic bandgap (PBG) within a certain range. Additionally, we may also take into account the ideal filtered wavelength at the output end of the fiber. Hence, the initial adjustment of the transmission signal can be achieved by regulating the angle at which it strikes. The behavior mentioned above is depicted in Fig. 8.3. The quantitative analysis shows that the suggested hollow-core high-contrast BFs optical inline filter can adjust its photonic bandgap (PBG) up to 828 nm by altering the incidence angle to 800. During this alteration in the angle at which the incident light strikes, the transmission signal that corresponds to resonance is likewise moved towards the blue end of the spectrum by 243 nm. This shift may be further adjusted by applying direct current (DC) voltages to the cavity. A marginal discrepancy of around 3% is seen in addition to Fig. 8.3. The variation in intensity is caused by the fluctuation in radiation loss of leaky defect cavity modes at varying angles of incidence. The BFs structure exhibits robust optical mode confinement at a certain angle of incidence. By altering the angle of incidence, the resonant condition for mode confinement is modified, resulting in small fluctuations in the intensity of the optical spectra seen in Figs. 8.3 and 8.4. This feature is very distinctive and can be utilized for sensing purposes. Figure 8.4 demonstrates the secondary electrical adjustable characteristics of the BFs structure. The PMN-PT single crystal is subjected to a 20 V DC voltage ranging from −10 V to +10 V in the depicted figure. This voltage causes a wavelength shift of 129 nm, resulting in both blue and red shifts depending on the applied DC voltage and angle of incidence. The result obtained demonstrates that this device has a wide application range, extending up to 129 nm, with a minimal voltage shift of around 20 V at a certain angle of incidence. This finding holds great promise for the development of multiwavelength LASER sources that cover a broad range of wavelengths. Furthermore, it is evident from Fig. 8.4 that the full width at half maximum (FWHM) of the transmission signal remains consistent and does not impact the fluctuation of the direct current (DC) voltage. In addition, we can also compute some performance characteristics for the hollow-core high-contrast optical inline filter. Figure 8.3 clearly indicates that the tuning range, as determined by the PBG shift, spans from 1123 nm to 1951 nm. Additionally, when examining the transmission signal for primary tuning, we observed that the tuning occurred at around 243 nm within the tuning range of 1402 nm to 1645 nm. The proposed optical inline filter device exhibits a spectral shift of 129 nm in the transmission signal when a 20 V direct current (DC) voltage is applied. The sensitivity of this device is defined as the ratio of the spectral change in the transmission signal to the small change in DC voltage, which is approximately 0.645 nm/mV. To assess the authenticity and superiority of the acquired result, we may compare the sensitivity of the current optical inline filter structure with the sensitivity achieved in a previous experiment using a fiber-optic comb filter based on PMN-PT, as demonstrated by Zhang et al. [34]. The sensitivity achieved in the comb filter structure [34] (about 0.000002 nm/mV) is significantly lower than the sensitivity achieved in the current BFs optical inline filter device structure. Moreover, the suggested filter device exhibits enhanced reliability due to its minimal Full Width at Half Maximum (FWHM) value described earlier. This improvement in FWHM would enhance both the Detection Accuracy (D.A.) and Quality of Performance (Q.P.), since these performance parameters are inversely related to the FWHM of the transmitted signal. Furthermore, Fig. 8.5 displays the remarkable linearity of the doubly electrically tuned hollow-core high-contrast BFs optical inline filter throughout its entire tuning range. The adjustable electric voltage enables a seamless and rapid sweep across the entire spectrum of wavelengths in both directions, ensuring consistent reproducibility.

8.5 Future Scope and Experimental Realization

A bandpass optical filter based on antiresonant hollow-core fibers [35] has been explored in recent comparable scientific studies, which also lay the groundwork for the experimental realization of Bragg fibers. To make the devices, a piece of tubular hollow-core fiber is tapered down to a ratio lower than 0.5. The blue end of the spectrum is suppressed by sweeping the resonant bands induced by the thickness of the tube wall in the down- and up-transition regions of the taper, and the red end of the spectrum is filtered by taking advantage of the increased confinement loss at the taper waist, which is strongly proportional to the ratio of wavelength to core diameter. By adjusting the fiber geometry and taper profile, one can alter the passband's position and breadth according to these filter operating principles. With a passband insertion loss of at least 1.3 dB and lossband suppression of up to 40 dB, it is possible to create a bandpass filter with a width of 350 nm. Filters are expected to play a crucial role in all-hollow-core fiberized optical systems.

Additionally, photonic bandgap Bragg fibers with a hollow core have several uses in industrial sensing. The development of practical sensing instruments places a premium on spectral scalability of the transmission window and detection accuracy [36]. By adjusting the fiber diameters and bilayer thicknesses, the experimental results show that the locations of the fiber bandgaps can be scaled with respect to wavelength. Improving the sensor system's detection accuracy and sensitivity to different frequencies can be achieved by adjusting the Bragg fiber geometry. To validate the methodology, both theoretical analysis and experimental demonstrations were carried out. A sensitivity improvement of about 32% was achieved by developing and producing Bragg fibers with an optimized bilayer thickness contrast. This work experimentally achieves the greatest value for spectral sensitivity for Bragg fiber-based refractive index sensors, which is 1850 nm/RIU. Furthermore, research into the effect of temperature on sensor performance has shown that our Bragg fiber sensor, when used with aqueous solutions in the fiber core, has a temperature stability of only 45 pm/°C. Online monitoring of the refractive index/concentration of numerous industrial fluids, trace amounts of biomolecules, real-time detection of binding and affinity, study of kinetics, and enhanced accuracy can all be achieved with the presented fiber sensor because it inherently integrates optical detection with microfluidics.

Furthermore, a reflecting sensor that measures both strain and temperature simultaneously is built by sandwiching a handmade hollow-core Bragg fiber (HCBF) [37, 38] between two single-mode fibers. This sensor is both proposed and shown. The suggested HCBF-based sensor guarantees two-parameter sensing, in contrast to conventional Fabry-Pérot interferometer (FPI) sensors that can only accomplish one-parameter sensing with inevitability cross-correspondence to other parameters. The HCBF serves as both an FPI-inducing FPI spectrum pattern and a weak waveguide confining light-inducing periodic envelope in the reflection spectrum. The restricting spectrum envelope and the FPI spectrum fringe are affected differently by the distinct sensing processes that cause different temperature and strain sensitivity values (2.98 pm/°C, 19.4 pm/°C, 2.02 pm/με, −0.36 pm/με) in the HCBF-based reflective sensor. The proposed sensor is able to measure both strain and temperature simultaneously using a 2×2 matrix, according to the experimental results. This sensor can sense both strain and temperature in extreme environments thanks to its small size, ease of production, and inexpensive price.

8.6 Conclusions

An investigation is conducted on a high-contrast optical inline filter with double electrical tuning capabilities. This optical inline filter is designed for low-power multiwavelength LASER applications and is located near the PMMA cavity with a single-crystal ferroelectric PMN-PT. The transmission peak via this structure has been achieved with an efficiency of around 90% in the studied photonic bandgap (PBG) by controlling the angle of incidence and applying direct current (DC) voltages. Applying an external voltage to the PMN-PT single crystal is necessary in order to control the thickness of the cavity. As a result, the PBG structure may be altered accordingly. This system has a wide applicability range, extending up to 129 nm. This is accomplished by exploiting the reversible domain switching capabilities of PMN-PT layers, which necessitates only a small voltage alteration of approximately 20 V. Therefore, the optical inline filter developed by BFs has significant promise for a wide range of multiwavelength LASER applications, meeting the requirements of current scenarios such as optical instrument testing, optical communications, fiber sensor systems, optical signal processing, modern telecommunications, and the sensors multiplexing industry.

References

1. M.W. Maeda, J.S. Patel, D.A. Smith, C. Lin, M.A. Saifi, A.V. Lehman, An electronically tunable fiber laser with a liquid-crystal etalon filter as the wavelength-tuning element. IEEE Photon. Technol. Lett. **2**(11) (1990)
2. J.M. Sierra-Hernandez, R. Rojas-Laguna, E. Vargas-Rodriguez, J.M. Estudillo-Ayala, R.I. Mata-Chavez, D. Jauregui-Vazquez, J.C. Hernandez-Garcia, J.A. Andrade-Lucio, J.C. Gutierrez-Gutierrez, A tunable multi-wavelength laser based on a Mach–Zehnder interferometer with photonic crystal fiber. Laser Phys. **23**, 055105. 5pp (2013)
3. M. Nasir, M.N. Al-Mansoori, M.H. Abdul Rashid, et al., Multiwavelength Brillouin-erbium fiber laser incorporating a fiber Bragg grating filter. Laser Phys. **18**, 446 (2008)

4. C.H. Yeh, C.W. Chow, Single-longitudinal-mode erbium-doped fiber laser with novel scheme utilizing fiber Bragg grating inside a ring cavity. Laser Phys. **20**, 512–515 (2010)
5. W. Yang, Y. Liu, L. Xiao, Z. Yang, Wavelength-tunable erbium-doped fiber ring laser employing an acousto-optic filter. J. Lightwave Technol. **28**, 118–122 (2010)
6. R.K. Kim, Y. Han, Switchable multiple lasing oscillations in an erbium-doped fiber ring laser using a single stage of a Sagnac loop mirror. Appl. Phys. B Lasers Opt. **103**, 813–818 (2011)
7. A. Castillo-Guzman, J.E. Antonio-Lopez, R. Selvas-Aguilar, D.A. May-Arrioja, J. Estudillo-Ayala, P. Likamwa, Widely tunable erbium-doped fiber laser based on multimode interference effect. Opt. Express **18**, 591 (2010)
8. C. Tu, W. Guo, Y. Li, S. Zhang, F. Lu, Stable multiwavelength and passively mode-locked Yb-doped fiber laser based on nonlinear polarization rotation. Opt. Commun. **280**, 448 (2007)
9. S. Pan, C. Lou, Y. Gao, Multiwavelength erbium-doped fiber laser based on inhomogeneous loss mechanism by use of a highly nonlinear fiber and a Fabry-Pérot filter. Opt. Express **14**, 1113 (2006)
10. F.F. Zhong, Y. Xu, Y.J. Zhang, et al., Widely ultra-narrow linewidth 104 nm tunable all-fiber compact erbium-doped ring laser. Laser Phys. **21**, 219–221 (2011)
11. N.K. Chourasia, A. Srivastava, V. Kumar, R.K. Chourasia, Doubly electrically tuned cylindrical Bragg fiber waveguide inline optical filter for multiwavelength LASER applications. Mater. Today Commun. **25**, 101620 (2020)
12. R.K. Chourasia, S. Prasad, V. Singh, Voltage-tunable pass band in a cylindrical multilayered structure containing PMMA and 0.67PMN–0.33PT single crystal as a defect layer. Opt. Quant. Electron. **48**, 539 (2016)
13. R.K. Chourasia, V. Singh, *Comparative analysis of the dispersion characteristic of Bragg fibers having different refractive index contrast*, in International Conference on Fibre Optics and Photonics, Th3A. 75 (2016)
14. R.K. Chourasia, C.S. Yadav, A. Upadhyay, N.K. Chourasia, V. Singh, Analysis of Bragg fiber waveguide having a defect layer for biosensing application. Optik **200**, 163400 (2020)
15. R.K. Chourasia, V. Singh, *Cylindrical Bragg waveguide-based temperature sensors*, in IEEE Uttar Pradesh Section International Conference on Electrical, Computer and Electronics Engineering (UPCON) 87-90 (2016)
16. R.K. Chourasia, C.S. Yadav, A. Upadhyay, N.K. Chourasia, V. Singh, Comparative study of defect mode intensity and wavelength modulation in Bragg fiber waveguide sensors. Optik **204**, 164198 (2020)
17. N.K. Chourasia, A. Srivastava, V. Kumar, R.K. Chourasia, Comparative performance study of liquid core cylindrical Bragg fibre waveguide biosensors. Pramana J. Phys. **95**, 9 (2021)
18. N.K. Chourasia, A. Srivastava, V. Kumar, R.K. Chourasia, Optimizing temperature-dependent molar volume fraction of biodiesel fuel in a pseudo-binary mixture through Bragg fiber waveguide sensor having defect layer. Fuel **293**, 120489 (2021)
19. R.K. Chourasia, N.K. Chourasia, A. Srivastava, N. Bihari, Photonic nanostructured Bragg fuel adulteration sensor, in *Photonic Materials: Recent Advances and Emerging Applications*, vol. 1, (2023), p. 237. https://doi.org/10.2174/97898150497561230100015
20. N.K. Chourasia, N. Bihari, R.K. Chourasia, An optofluidic Bragg fiber sensor for estimating adulterants in a temperature-dependent molar fraction of hydrated mono-alcohol fuels. Heliyon **8**(9), e10532 (2022)
21. N. Bihari, A. Srivastava, N.K. Chourasia, R.K. Chourasia, H-polarised EM-wave transport in polymeric-chalcogenide columnar photonic materials. J. Phys. Conf. Ser. **2426**, 012009 (2023)
22. R.K. Chourasia, N.K. Chourasia, N. Bihari, Optical properties of hollow-core Bragg fiber waveguides, in *Photonic Materials: Recent Advances and Emerging Applications*, vol. 1, (2023), p. 214. https://doi.org/10.2174/9789815049756123010014
23. N. Bihari, N.K. Chourasia, R.K. Chourasia, Exploring optical properties in cylindrical polymeric-chalcogenides photonic materials. Mater. Today Proc. **67**, 625–631 (2022)
24. R. Zhang, B. Jiang, W. Cao, Elastic, piezoelectric, and dielectric properties of multidomain $0.67Pb(Mg_{1/3}Nb_{2/3}O_3\text{-}0.33PbTiO_3)$single crystals. J. Appl. Phys. **90**, 3471–3475 (2001)

25. R. Zhang, B. Jiang, W.H. Jiang, W. Cao, Complete set of elastic, dielectric, and piezoelectric coefficients of $0.93Pb(Zn_{1/3}Nb_{2/3})O_3$-$0.07PbTiO_3$ single crystal poled along [011]. Appl. Phys. Lett. **89**, 242908–242903 (2006)
26. E.W. Sun, S.J. Zhang, J. Luo, T.R. Shrout, W. Cao, Elastic, dielectric, and piezoelectric constants of Pb,In1/2Nb1/2…O3–Pb,Mg1/3Nb2/3…O3–PbTiO3 single crystal poled along [011]c. Appl. Phys. Lett. **97**, 032902–032903 (2010)
27. M.A. Kaliteevski, R.A. Abram, V.V. Nikolaev, G.S. Sokolovski, Bragg reflectors for cylindrical waves. J. Mod. Opt. **46**, 875–890 (1999)
28. A.H. Cherin, *An Introduction to Optical Fibers* (McGraw-Hill, Tokyo, 1983)
29. M. Born, E. Wolf, *Principles of Optics* (Cambridge University Press, Cambridge/London, 1999)
30. A.W. Snyder, J. Love, *Optical Waveguide Theory*, 2nd edn. (Springer, 2008)
31. R.K. Chourasia, V. Singh, Estimation of the photonic band gap in the hollow core cylindrical multilayer structure. Superlattice. Microst. **116**, 191–199 (2018)
32. R.K. Chourasia, S. Prasad, V. Singh, Performance study of a liquid-core Bragg fiber sensor in the presence of a defect layer. Opto-Electron. Rev. **25**, 215–221 (2017)
33. A.F. Abouraddy et al., Towards multimaterial and multifunctional fibers that see, hear, sense and communicate. Nat. Mater. **6** (2007). https://doi.org/10.1038/nmat1889
34. F. Zhang, X.F. Jin, X.M. Zhang, S.L. Zheng, H. Chi, K.Y. Zou, J.P. Yao, Fibre-optic comb filter with tunable central wavelength and channel isolation at sub-microsecond speed. Electron. Lett. **46**, 71–73 (2010)
35. D. Xiong, X. Wu, M.R.A. Hassan, T. Gavara, W. Chang, In-line hollow-core fiber-optic bandpass filter. Opt. Lett. **46**, 5918–5921 (2021)
36. J. Li, K. Nallappan, Optimization of hollow-core photonic Bragg fibers towards practical sensing implementations. Opt. Mater. Express **9**, 1640–1653 (2019)
37. Y. Wang, Y. Zhou, X. Wang, D. Chen, Z. Lian, C. Lu, H.-Y. Tam, Simultaneous measurement of temperature and strain based on a hollow core Bragg fiber. Opt. Lett. **45**, 6122–6125 (2020)
38. B. Huang, X. Sheng, J. Cao, S. Lou, A compact sensor integrated in a single hollow core Bragg fiber for simultaneous measurement of temperature and strain. IEEE Sensors **23**(10), 10478–10488 (2023)

Chapter 9
Bragg Fiber: Some Nonlinear Aspects

Abstract This chapter briefly discusses the various recent Bragg fiber nonlinear characteristics that are very relevant to photonics applications. First, an attempt is made to examine solitons in a dispersive ring that shows Kerr nonlinearity and experiences no loss. The nonlinear transmission of optical pulses in solid-core Bragg fibers with durations of several hundred femtoseconds is then briefly summarized. Subsequently, the development of sech pulses in the nonlinear domain was investigated. The investigation focusses on the influence that pulse width, peak power, soliton order, and higher order dispersion have on the development of each pulse. Next, an intriguing demonstration of the propagation of megawatt-class pulses in flexible all-silica fibers has been investigated. It is now known that well designed all-glass antiresonant fibers, when pumped by fiber-based chirped pulse amplification (CPA) devices, make it possible to generate and distribute megawatt-class pulses over extensive distances in an effective manner. These specific types of Bragg fibers have the potential to facilitate the creation of miniaturized endomicroscopes that can be utilized for a variety of biomedical purposes. Finally, a novel tellurite high index core Bragg fiber has been studied as a novel method for generating a powerful supercontinuum. High-index-core Bragg fibers have significant potential for achieving extremely coherent mid-infrared supercontinuum production.

9.1 Introduction

Annular photonic structures, found in both bulk and fibers, are Bragg structures that exhibit both Bragg resonances and waveguiding properties. Bragg fibers are optical waveguides consisting of a central region with a low refractive index, serving as the core, that is enveloped by concentric layers of alternating materials with high and low refractive indices. The periodicity of the refractive index in the cladding gives rise to a photonic bandgap (PBG) [1, 2]. The incident light may be confined to the fiber's core if its frequency or wavelength falls inside the bandgap. Therefore, Bragg fibers are alternatively referred to as 1-D PBG fibers. They are associated with photonic crystals due to their significant disparity in refractive indices $(n_2 - n_1)$,

R. K. Chourasia, A. Katti, *Bragg Fibers*,
https://doi.org/10.1007/978-3-031-65164-9_9

which necessitates the consideration of the reflected wave in its whole, surpassing the approximations often used for analytical gap solitons. Their ability to confine is crucial, but the potential of these structures is significant and warrants additional investigation. Waveguiding happens in a homogeneous structure and can be managed using established methods.

The fiber exhibits a nonlinear response to the propagating light when the light intensity traversed through it reaches a level where its associated electric field aligns with the local electric field of the atomic dipoles. The induced polarization in the fiber is mathematically represented as having a nonlinear dependence on the electric field of the traveling beam [3] to depict this nonlinear optical response. The behavior of a pulse that is spreading out in this region with nonlinear properties is affected by both its strength and the dispersion properties of the fiber. In order to establish effective nonlinear interactions in an optical fiber, it is typically necessary to operate at a wavelength where the group velocity dispersion (GVD) is almost equal to zero. In recent years, the advancement of microstructured fibers such as photonic crystal and photonic bandgap (PBG) fibers has made it possible to achieve zero group velocity dispersion (GVD) at wavelengths significantly shorter than the typical 1310 nm wavelength in traditional silica fibers. This provides a wide spectrum of optical frequencies for investigating nonlinear optical phenomena in optical fibers, namely in the visible and near-infrared areas [4]. Few studies have detailed the distribution of solitons in a nonlinear circular and annular region [4, 5]. While the predominant body of research on Bragg fibers centers on air-core configurations that preclude nonlinear interactions, investigations have been carried out on Bragg fibers featuring solid cores, which do facilitate nonlinear propagation [4]. Silica, by virtue of its finite Kerr nonlinearity and third-order susceptibility, may be examined as a prospective substitute substrate in the investigation of nonlinear pulse propagation within a solid-core Bragg fiber. This provides a substitute for nonlinear interactions, like obtaining supercontinuum light, utilizing photonic crystal fibers. This chapter will initially examine the fundamentals of optical solitons on circular and annular structures through an analysis of the stable states resulting from the interaction of solitons in a ring configuration. Next, we will examine how the input pulse parameters and the properties of the Bragg fiber affect the creation of small, powerful optical soliton pulses as they travel through solid-core Bragg fibers in the nonlinear optical region.

9.2 Solitons in a Lossless Nonlinear Ring

First, let's examine a dispersive ring that does not experience any loss and demonstrates a Kerr nonlinearity [5]. The system is governed by the nonlinear Schrodinger equation (NLSE) with periodic boundary conditions. This equation has closed form solutions in the form of the Jacobian elliptic function [6], which can be obtained through the inverse scattering transform. These solutions pertain to the occurrence of periodic waves that span the entire ring. However, we shall see the stable states of individual solitons that are separately created within the ring and experience a gradual change in their relative position. Therefore, we start with a total of identical basic solitons, which weakly interact with each other due to their partial overlap. The equation u_n describing the envelope of the nth soliton in the ring is the Nonlinear Schrödinger Equation (NLSE) modified by the influence of surrounding solitons [5],

$$\frac{\partial u_n}{\partial z} - jD\frac{\partial^2 u_n}{\partial t^2} - j\delta|u_n|^2 u_n = j\delta\left[2|u_n|^2 u_{n-1} + u_n^2 u*_{n-1} + 2|u_n|^2 u_{n+1} + u_n^2 u*_{n+1}\right] \quad (9.1)$$

retaining solely the initial terms in the perturbation. In this context, z represents the distance traveled by a wave along the ring, whereas t represents the time measured in a coordinate system that is fixed to an individual soliton that does not interact with other solitons.

$D = -\beta''/2$ is the group velocity dispersion, where β'' is the second derivative of the propagation constant, and δ is the Kerr parameter. The system of N equations is subject to a cyclic boundary condition $u_n = u_{n+N}$, which is crucial for achieving equilibrium states as it ensures that all solitons undergo comparable interactions. Under the assumption of a minor perturbation $\Delta u_n(z,t)$ caused by the adjacent solitons, the nth soliton envelope can be expressed as,

$$u_n(z,t) = \left[U_n^0(t) + \Delta u_n(z,t)\right]e^{j\left(-Dp_n^2 z + Dz/\tau^2\right)} \quad (9.2a)$$

$$U_n^0 = A_0 \sec h\left(\frac{t-t_n}{\tau}\right)e^{jp_n t + j\theta_n} \quad (9.2b)$$

where A_0 is the amplitude of the soliton, τ is its width, p_n is the divergence of the nth soliton center frequency from the expected frequency ω_0, and θ_n is the phase of the nth soliton. Equation (9.2b) represents the unperturbed soliton. By employing a linearized soliton perturbation theory [7], we can solve the equations of motion that describe the soliton's energy Δw_n, frequency Δp_n, phase $\Delta\theta_n$, and timing Δt_n,

$$\frac{\partial}{\partial z}\left(\frac{\Delta w_n}{w_0}\right) = \delta A_0^2.4A_0\left[e^{-\frac{t_{n+1}-t_n}{\tau}}\sin(\theta_{n+1}-\theta_n) - e^{-\frac{t_n-t_{n-1}}{\tau}}\sin(\theta_n-\theta_{n-1})\right] \quad (9.3a)$$

$$\frac{\partial}{\partial z}(\Delta p_n.\tau) = \delta A_0^2.(-4A_0)\left[e^{-\frac{t_{n+1}-t_n}{\tau}}\cos(\theta_{n+1}-\theta_n) - e^{-\frac{t_n-t_{n-1}}{\tau}}\cos(\theta_n-\theta_{n-1})\right] \quad (9.3b)$$

$$\frac{\partial}{\partial z}(\Delta\theta_n) = \delta A_0^2\left(\frac{\Delta w_n}{w_0}\right) + \delta A_0^2.(-6A_0)\left[e^{-\frac{t_{n+1}-t_n}{\tau}}\cos(\theta_{n+1}-\theta_n) - e^{-\frac{t_n-t_{n-1}}{\tau}}\cos(\theta_n-\theta_{n-1})\right] \quad (9.3c)$$

$$\frac{\partial}{\partial z}\left(\frac{\Delta t_n}{\tau}\right) = \frac{2d}{\tau^2}(-\Delta p_n.\tau) + \delta A_0^2.2A_0\left[e^{-\frac{t_{n+1}-t_n}{\tau}}\sin(\theta_{n+1}-\theta_n) - e^{-\frac{t_n-t_{n-1}}{\tau}}\sin(\theta_n-\theta_{n-1})\right] \quad (9.3d)$$

where $w_0 = 2A_0^2\tau$ is the energy of the unperturbed soliton.

The equilibrium state is achieved when the soliton's energy and frequency fluctuations are time-independent, meaning that Eqs. (9.3a and 9.3b) are equal to zero, producing,

$$\theta_{n+1} - \theta_n = \theta_n - \theta_{n-1} = \phi \tag{9.4a}$$

$$t_{n+1} - t_n = t_n - t_{n-1} = T \tag{9.4b}$$

Consequently, it can be deduced that the perturbations in both the timing difference and phase difference must be zero, as indicated by Eqs. (9.3c and 9.3d),

$$\Delta w_{n+1} = \Delta w_n \tag{9.5a}$$

$$\Delta p_{n+1} = \Delta p_n \tag{9.5b}$$

These two requirements are easily met due to the symmetry of the equilibrium state. Hence, at a state of equilibrium, the solitons are distributed uniformly throughout the ring with consistent timing and phase disparities, in accordance with the system's symmetry. The region $\pi/2 < \phi < 3\pi/2$ is emphasized due to the requirement for stable equilibrium state, which can only be achieved when the interaction between solitons is repulsive. The phase difference between solitons must conform to a specific format due to the periodic boundary requirement $\theta_{N+1} = \theta_1 + N$. $\phi = \theta_1 + m.2\pi$,

$$\phi = \frac{2m}{N}\pi \tag{9.6}$$

When an even number of solitons are present in a ring, the equilibrium state will have a different configuration than when an odd number of solitons are present in the ring. This differs from the research carried out in reference [8], which specifically concentrated on investigating the equilibrium states of two solitons within a ring. As a result, it could not forecast the stable conditions of an odd quantity of solitons. Another significant finding is that the precise time of the solitons in relation to the ring coordinate system may vary. After setting $\Delta p_n = 0$, in (8.3d) (The simulation results validate that the center frequency remains unaltered by the interaction between solitons in a state of equilibrium), we find, in equilibrium,

$$\frac{\partial}{\partial z}\left(\frac{\Delta t_n}{\tau}\right) = 4\delta A_0^3 \sin\phi . e^{-\left(\frac{T}{\tau}\right)} \tag{9.7}$$

When $\phi \neq \pi$, e.g., for an odd N, we obtain $\frac{\partial}{\partial z}(\Delta t_n) \neq 0$. In the above scenario, it is observed that all solitons experience concurrent motion, resulting in changes to their absolute timing with respect to the ring coordinate system. However, the solitons preserve their equilibrium condition, as evidenced by the absence of any alterations in their relative timing and phase difference. The verification of the analytical conclusions' validity was conducted by the utilization of a split-step fast Fourier transform (FFT) approach to represent the movement of solitons in a ring [9]. To integrate the periodic boundary conditions of the ring, a discrete Fast Fourier Transform (FFT) was used with the same periodicity as the ring. There is a specific stable condition of an equal number of solitons in a ring, with some solitons having an even number and others having an odd number [5]. The soliton train's motion in relation to the ring system, where there is an odd number of solitons, can be observed through the inclined path on the graph. This motion aligns with Eq. (9.7). The equilibrium states we see for both even and odd numbers of solitons in a lossless ring are, nevertheless, unstable. From observation (9.3), it is evident that the properties of each soliton, such as energy, frequency, timing, or phase, are interconnected with the parameters of the adjacent solitons. If any of the parameters of the soliton deviate from equilibrium, it will cause oscillations in all of the soliton parameters. These oscillations will then spread to additional solitons along the ring. Given that the aforementioned system lacks dissipation, the oscillations will endure without any dampening effect, resulting in the system's inability to reach a state of equilibrium. Due to the intricate nature of the interconnected equations that describe the soliton system, the way they develop relies on the precise initial conditions, and it is not possible to examine the overall scenario using analytical methods.

9.3 Nonlinear Pulse Propagation Through Bragg Fiber

As previously stated, the finite Kerr nonlinearity of silica, specifically its third-order susceptibility, provides a reason to investigate the nonlinear propagation of pulses in a solid-core Bragg fiber. This fiber is considered as a potential alternative platform for nonlinear interactions, instead of using photonic crystal fibers. This motivation is additionally supported by the fact that the fabrication process may be easily accomplished using the advanced and well-established technology of modified chemical vapor deposition (MCVD), ensuring consistent and replicable results. This section will examine how input pulse parameters and fiber characteristics affect the development of brief, high-intensity optical pulses as they travel through solid-core Bragg fibers in the nonlinear optical zone.

Previously, we have observed that the analysis of Bragg fibers can be conducted using the transfer matrix theory. The analysis of pulse propagation in Bragg fibers can be reduced using the transfer matrix approach, especially when dealing with low-index contrast fibers. This simplification allows for the utilization of the widely recognized scalar linearly polarized (LP) mode approximation. Considering the inherent tendency of Bragg fiber modes to leak, an analytical technique known as the matrix approach [10] can be employed to study and evaluate these modes. The matrix approach stated above was initially suggested to investigate the leaky modes of a conventional silica fiber. Under the LP approximation, the radial portion of the modal field at different layers of the Bragg fiber is expressed using the following form [4].

$$R(r,z,\theta,t) = \left[A_j J_l(k_j r) + B_j Y_l(k_j r)\right]\left[\cos(l'\theta)\sin(l'\theta)\right]\exp[i(\omega t - \beta z)] \quad (9.8)$$
$$j = 1, 2 \ldots, Q, l, l' = 0, 1, 2, \ldots$$

where $k_j^2 = n_j^2(\omega^2/c^2) - \beta^2$, n_j represents the refractive index of the jth layer, the symbol β represents the propagation constant, and Q represents the total number of cladding layers. To analyze each mode with the indices l (representing the order of the Bessel functions) and l', we apply the continuity conditions for the electric field and its derivatives at the interfaces of each layer in the Bragg fiber. This allows us to establish relationships between the coefficients of the field in each layer and those of the neighboring layer using 2×2 matrices [11].

In order to study the formation of a short and powerful optical pulse, we analyze a Bragg fiber consisting of a silica core surrounded by 10 layers of GeO_2-doped silica. The cladding's alternating layers with high and low refractive indices can be achieved by adjusting the doping level of GeO_2, as mentioned above. The fiber's index profile characteristics were selected to ensure that the group velocity dispersion (GVD) at 1060 nm was close to zero. The Bragg fiber we have considered has a silica core with a radius of 5.82 μm. The cladding layers surrounding the core have alternating thicknesses of 1.36 μm and 1.55 μm. The cladding layers have refractive indices of 1.47 and 1.454, respectively, at a wavelength of 1060 nm. The fiber has a positive net temporal dispersion of +6.4 ps/(nm · km) at a wavelength of 1060 nm [4]. It should be emphasized that standard telecommunications-grade silica fibers are incapable of exhibiting anomalous dispersion characteristics at wavelengths of 1060 nm or below. The fiber's mode effective area, as determined in reference [9], was approximately 55 μm^2.

The progression of an optical pulse in the single-mode domain of the fiber is determined by [9].

$$\frac{\partial A}{\partial z} + \frac{\alpha}{2}A + \beta_1\frac{\partial A}{\partial t} + i\frac{\beta_2}{2}\frac{\partial^2 A}{\partial t^2} - \frac{\beta_3}{6}\frac{\partial^3 A}{\partial t^3} + \ldots = i\gamma\left[1 + \frac{i}{\omega_0}\frac{\partial}{\partial t}\right]$$
$$\times \left[A(z,t)\int_{-\infty}^{\infty} R(t')|A(z,t-t\prime)|^2 dt'\right] \quad (9.9)$$

$R(t')$, or the Raman response function incorporates both Raman contributions, i.e., the vibrational and electrical part. $A(z, t)$ denotes the gradually varying magnitude of the propagating mode. α represents the attenuation of signal propagation. β_n's are the dispersion coefficients corresponding to various orders. γ represents the effective nonlinear coefficient of the fiber. The Raman response function, denoted as $R(t')$, encompasses both the electrical and vibrational Raman contributions. To solve (9.2), the split-step Fourier technique incorporating dispersion terms up to the fifth order is utilized [9]. Over 200 data points were utilized during a time interval that was 60 times greater than the duration of the pulse. A theoretical analysis on the impact of dispersion on the development of an input sech pulse with a full width at half maximum (FWHM) of 100 femtoseconds, centered at a wavelength of 1060 nanometers was conducted. The pulse's peak power (P) was modified, leading to fluctuations in the soliton-like pulse's magnitude. A soliton with an order greater than 1 (order >1) undergoes regular compression as it propagates through the fiber in the phase of anomalous dispersion. Through the simulation of soliton pulse propagation at various orders, we see that the pulse is maximally compressed at a distance approximately equal to $z_0/(N - 1)$, where z_0 represents the soliton period and N is an integer value greater than or equal to 2. After surpassing this distance of transmission, the pulse starts to disintegrate, and the accompanying frequency spectrum does not display any additional widening [4, 12]. The pulse is characterized as "soliton-like" and not exactly a "soliton" because it incorporates higher order dispersion terms and mathematical expressions denoting the Raman and self-steepening effects. Consequently, Eq. (9.9) does not fully support a soliton in its strict definition. On the other hand, as the evolution continues, these factors cause the sech pulse to be perturbed, which would otherwise be a soliton at the input.

Subsequently, it can be examined how the soliton order impacts the frequency spectrum of the output pulse. In order to achieve a wider frequency spectrum, it has been proved that it is possible to achieve this by preserving the same soliton order while simultaneously increasing the peak power and the amplitude of the positive dispersion. As a result, it is possible to conduct an analysis of the development of the pulse as it travels through the fiber. The pulse spectrum can be investigated in both the temporal and frequency domains as the pulse travels through the suggested Bragg fiber. The fiber has a dispersion coefficient (D) of +6.4 ps/(nm · km) at a wavelength of 1060 nm. It can be shown that the impacts of self-phase modulation (SPM) and group velocity dispersion (GVD) on the higher order soliton have the greatest impact on the initial widening [4]. The asymmetry in the frequency spectrum is caused by the significant third-order dispersion ($\beta_3 = 0.0301$ ps^3/km) of the fiber. A peak that is roughly 72 nm from the pulse's center emerges at a shorter wavelength as the pulse moves [4].

The expansion of the pulse's spectrum was effectively restricted to less than 100 nm due to the substantial dispersion slope of the Bragg fiber that was mentioned. In order to provide a broader output spectral range, attempts have been made to achieve a convex dispersion profile. The presence of convex dispersion profiles in optical fibers has been demonstrated to lead to increased nonlinear interactions [13]. Dispersion management techniques, such as utilizing tapered conventional or holey fibers to decrease dispersion or regulate it, have been recognized as effective approaches to broaden the frequency spectrum of fiber output for short, high-intensity input pulses. The dispersion decreasing Bragg fibers (DDBF) are created by systematically increasing the fiber's diameter during the drawing process. This leads to variations in both the structural dimensions and the dispersion of the fiber along its length. The dispersion characteristics of the tapered Bragg fiber are significantly influenced by its physical parameters. Consequently, the dispersion characteristics vary across different taper ratios of the fiber [14, 15]. The taper ratio refers to the ratio between the core diameters of the Bragg fiber at the output end and the input end. The thickness of the cladding layer of the fiber is also proportionally adjusted. A DDBF has been built and pulse propagation has been investigated through it, taking into account the features indicated above [4]. Through the examination of propagation dynamics, it was noted that the higher order soliton pulse eventually undergoes disintegration. While the propagating pulse's frequency spectrum expands due to the frequency components of each fundamental soliton. Simultaneously, dispersive waves are generated, which begin to spread away from the original pulse, leading to a significant loss of energy. Subsequently, as the complete pulse falls inside the typical dispersion range, it experiences fragmentation, resulting in the cessation of further widening.

9.4 Megawatt Soliton Generation in Bragg Fiber

Low-index core antiresonant fibers have a key characteristic of radiation leakage, which results in a strict mode selection and enables single-mode operation, even with a large core area [16]. The broad fundamental bandgap of silica-core Bragg fibers enables significant frequency changes through the soliton self-frequency shift phenomenon. It has been shown that well-designed all-glass antiresonant fibers, when pumped by fiber-based chirped pulse amplification (CPA) devices, enable the efficient creation and delivery of megawatt-class pulses over large distances [17].

Following the Ref. [17], let us consider a fiber preform created using the modified chemical vapor deposition (MCVD) technique. The refractive index profile was determined using the refracted near field approach with the NR9200 device from EXFO. The core has a diameter of 80 μm. The index contrast of the high-index layers is 0.018. The index contrast of the outer high-index layer seems reduced because it is challenging to reliably quantify such index contrasts at large radii using the refracted near-field technique. The high-index layers have an average thickness of around 2 μm, whereas the low-index layer is 10 μm thick. The alignment between the measured index profile and the targeted one is exceptional, considering the intricate nature of the index profile that includes very abrupt index changes. The fundamental mode's effective area was calculated to be 2710 μm^2 at a wavelength of 2000 nm, making it the greatest effective area ever recorded for a flexible fiber. We have considered a bespoke master-oscillator power-amplifier system using Thulium-doped large mode area fiber to showcase exceptional energy levels. The laser system emits 700 femtosecond pulses at a wavelength of 1970 nanometers, with a maximum energy of 10 microjoules, and operates at a repetition rate of 400 kilohertz. The performances were fine-tuned to improve soliton generation in short segments of the Bragg fiber with a very large mode area. The pulse train was transmitted into the antiresonant fiber using a long focal length $f_1 = 50$ mm lens with a maximum launching effectiveness of 55%. The beam was focused before analysis using a $f_2 = 8$ mm lens. The findings are outlined in Fig. 9.1. Figure 9.1a displays the seed laser spectrum and the beam intensity distribution at the antiresonant fiber output during low-power operation. The seed laser is directed into the Gaussian-like mode of the antiresonant fiber, with minimal light in the high-order modes of the high-

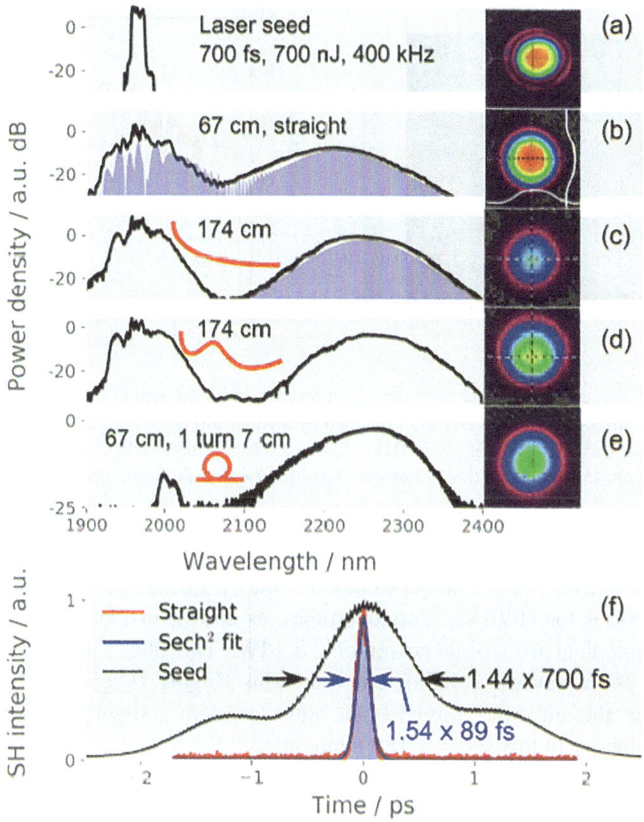

Fig. 9.1 Propagation of solitons in a flexible antiresonant fiber initiated at 1970 nm. Top: Configuration for experimentation. Middle: Spectra were measured for the (**a**) seed laser, (**b**) the straight fiber, (**c**) a huge bend, (**d**) loose coils, and (**e**) a bend with a radius of 7 cm. Insets depict the distributions of near-field intensity. Scales vary. Numerical findings are displayed in the filled areas in (**b**), and a sech^2 fit of the soliton spectrum is shown in (**c**). Autocorrelation traces have been determined for the seed laser and the solitonic pulse at 2200 nm in the straight fiber. Ref. [17]. (Reprinted with permission from H. Delahaye, G. Granger, D. Gaponov, L. Lavoute, S. Aleshkina, M. Salganskii, A. Hideur, M. Likhachev, and S. Février, "Megawatt solitons generated above 2000 nm in Bragg fibers," Opt. Lett. 44, 2713–2715 (2019) © 2019 Optical Society of America)

index rings, as illustrated in the inset of Fig. 9.1a. The input power was adjusted to facilitate the creation of solitons and self-frequency shift in a 67 cm long straight fiber. Figure 9.1b displays the output spectrum with a soliton formed at 2220 nm and a bandwidth of 95 nm full width at half maximum (FWHM). The inset demonstrates that the near field intensity distribution of the soliton is circular and does not extend into the cladding rings. A soliton with a wavelength of 2220 nm was separated using a bandpass filter with a bandwidth of 500 nm. The pulse energy in this setup is 252 nanojoules.

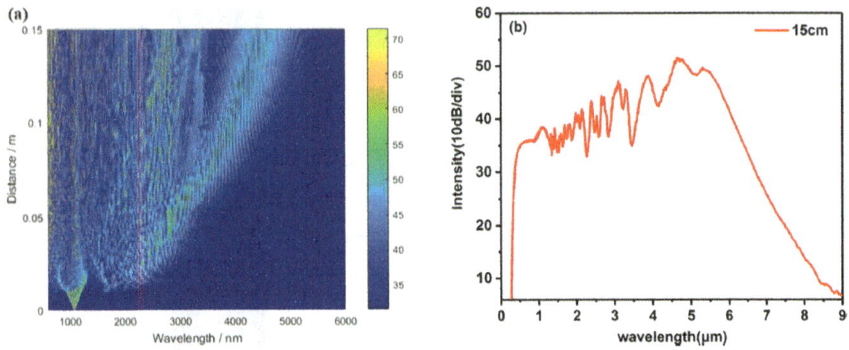

Fig. 9.2 (**a**) The changes in SC over time as it travels down the length of propagation, and (**b**) spectra of SC after propagation of 15 cm pumped by a 1.064 μm pulse with a 100 fs pulse duration and a peak power of 2 kW, for $dn = 0.02$, $\Lambda = 1.2$, $R = 5$, $d/\Lambda = 0.15$ [29]. (Reprinted with permission from Publisher, Taylor & Francis Ltd., http://www.tandfonline.com: [29])

The full width at half maximum (FWHM) pulse duration of 89 femtoseconds was determined from the autocorrelation trace using a second-harmonic (SH) autocorrelator (130-XL, Femtochrome), as shown in Fig. 9.1f. This energy level is equivalent to a peak power of 2.8 MW. The fiber's behavior has been examined under several bending configurations. Figure 9.2c displays a soliton generated in the smoothly curved fiber, however, with a slightly bigger red-shift. The pulse energy in this setup is 189 nanojoules.

This is an appreciable amount of energy for an ultrashort pulse produced and transmitted through a curved solid-core fiber. The somewhat longer center wavelength is attributed to the decreased effective area. The fiber was loosely coiled on the table as depicted in Fig. 9.1d. This specific arrangement was selected for analysis due to its proximity to the actual scenario for electricity distribution. The spectrum decreases significantly, but the pulse energy remains strong at 155 nJ. For a tight bending radius of 7 cm (Fig. 9.1e), a soliton developed and experienced self-frequency shift. The soliton energy recorded in this example was 147 nanojoules. The bandgap of the fiber decreases when tightly bent. When tightly bent, the fiber filters out residual pump at 1970 nm, as depicted in Fig. 9.1e. It is anticipated that the decrease in the modal effective area caused by the bend, along with the resulting rise in the effective nonlinear coefficient [18], could counteract the loss caused by the bend and support the circumstances necessary for high-energy soliton generation. Based on the spectrum analysis, we deduce that the transform-limited solitonic pulse lasts around 120 femtoseconds for the setups depicted in Fig. 9.2c, d. This confirms the generation and propagation of megawatt solitons across extended distances in slightly curved fibers.

The above study demonstrates the propagation of megawatt-class pulses in flexible all-silica fibers [17]. A device utilizing Erbium-doped fiber chirped-pulse amplification at 1560 nm could generate megawatt peak power pulses in the same fiber. The fibers could enable the development of miniaturized endomicroscopes operating at a wavelength of roughly 1700 nm, where high-power sub-100 fs pulses are needed for three-photon stimulated fluorescence microscopy using red fluorescent proteins.

The spectral region near 1700 nm is ideal for detailed multiphoton imaging of intact tissues because of the decreased scattering and absorption at this wavelength. Yet, widespread use of this excitation range will depend on the presence of durable and affordable high peak power pulsed lasers functioning at these wavelengths. Previous investigations have shown a fiber-based femtosecond laser producing pulses of 95 nJ and 85 fs duration at a wavelength of 1800 nm. The laser system utilizes a fiber-based chirped pulse amplifier that emits light at 1560 nm, and is then connected to a very large mode area antiresonant fiber that is produced in-house to achieve soliton self-frequency shift. The flexible fiber provides direct access to megawatt-peak power pulses at a repetition rate of 1 MHz. The source exhibits significant potential for biological microscopy by the acquisition of three-photon-excited fluorescence pictures of mouse nervous tissue [19]. The utilization of a flexible fiber specifically designed to transmit megawatt solitons within the biologically significant range of around 1700 nm presents an opportunity for conducting deep brain imaging of animals in motion using miniaturized endomicroscopes. The first study conducted in [19] focuses on the creation of megawatt peak-power pulses within the biologically significant range of 1700 nm using a long and flexible single-mode fiber. A flexible antiresonant optical fiber of the Bragg type has been developed to enable single mode propagation and resistance to bending, with an effective area exceeding 2000 μm^2 at a wavelength of 1560 nm. A large core fiber, when stimulated by short pulses from an Erbium-doped fiber-based Chirped Pulse Amplification system, produces and transmits solitonic pulses with energy of up to 95 nJ and duration of 85 fs at distant wavelengths of up to 1800 nm. The potential of the source by imaging the mouse brain ex vivo has been demonstrated [19]. During the imaging studies, 8 mW was applied to the sample from a source capable of delivering up to 70 mW at 1680 nm. Therefore, a ninefold increase in power at the focus may be anticipated. This restriction is caused by the sample surface overheating at higher power levels, which is especially significant in our work with a stationary sample compared to imaging living tissue under comparable circumstances [20]. Utilizing a custom objective with improved transmission and field of view, along with optimized brain labelling, could significantly increase the potential imaging depth beyond the existing limitation set by the laser source. Deep brain imaging of freely moving animals using a multiphoton endomicroscope based on this fiber could be considered.

9.5 Supercontinuum Generation in Bragg Fibers

The mid-infrared supercontinuum source, (MIR SC), has garnered significant interest for its wide range of applications in sectors like spectroscopy [21, 22], biomedical technology [23, 24], and gas remote sensing [25]. Chalcogenide step-index optical fibers and microstructured photonic crystal fibers have been successfully created by SC in the last 10 years [26, 27]. Single-mode fiber has a singular structure with limited design options. Preparing PCFs is challenging due to their intricate nature. Several novel optical fibers, including high-index-core (HIC) Bragg fiber, have been suggested for generating MIR SC. In the mid-infrared (mid-IR) band, the HIC Bragg fiber exhibits dispersion properties which can be controlled and a strong nonlinear coefficient. It is possible to alter the chromatic dispersion profile as well as the position of the zero dispersion point by adjusting the core diameter and the cladding thickness of the HIC Bragg fiber [28].

Reference [29] provides a description of an HIC Bragg fiber featuring a tellurite core with a diameter of 5 μm. The dispersion is extremely flat and has a zero dispersion point at 1.064 μm. The HIC Bragg fiber, with a zero dispersion wavelength of around 1 micron, is intended to be suitable for use with durable, mode-locked Yb-fiber lasers for femtosecond laser applications [30]. This compatibility theoretically simplifies the generation of mid-infrared supercontinuum. The investigation focused on the generation of SC spectra through the manipulation of fiber length and pump wavelength. The fiber, measuring 15 cm in length, was subjected to a pumping process at a wavelength of 1.064 μm using a pump power of 2 kW. This process yielded a wide-ranging supercontinuum spectrum spanning from 0.6 to 7.8 μm. The HICBF is composed of a core with a high refractive index, which is encased by a cladding made up of alternating layers of tellurite and a dielectric with a low refractive index. The core radius is represented by R, the radial multilayer period is denoted by λ, and the low-index-layer thickness is d. The refractive indices alternated between n_1 and n_2, with n_2 being less than n_1. The 30TeO2-30GeO2-30PbO-10PbF2 tellurite glass was selected as the high refractive-index core material, while the 30TeO2-30GeO2-30PbO-5PbF2-5Nb2O5 was chosen as the lower refractive-index dielectric. At a wavelength of 1.064 αm, the material with the high refractive index exhibited a refractive index of approximately 2.04, while the material with the low refractive index displayed a refractive index of approximately 2.02. As a result of this design, the Bragg fiber possesses refractive indices of 2.04 and 2.02, a core radius of 5 μm, and cladding layers that alternate between 0.13 μm and 1.37 μm in thickness. The influence of chromatic dispersion on the characteristics of MIR SC is clarified through a coherence spectrum and numerical analysis of intensity. The calculation of MIR SC generation was performed with the generalized nonlinear Schrödinger equation (GNLSE) [29, 31],

$$\begin{aligned}\frac{\partial A}{\partial Z}+\frac{\partial}{2}A+\frac{i}{2}\beta_2\frac{\alpha^2 A}{\partial T^2}-\frac{1}{6}\beta_3\frac{\alpha^3 A}{\partial T^3}-\frac{i}{24}\beta_4\frac{\partial^4 A}{\partial T^4}\\ =i\gamma\left[|A|^2A+i\frac{\lambda_c}{2\pi c}\frac{\partial}{\partial T}\left(|A|^2A\right)-T_RA\frac{\partial|A|^2}{\partial T}\right]\end{aligned} \tag{9.10}$$

where the slowly varying propagating mode's complex amplitude is represented by A, λ_c is the center wavelength, βi ($i = 1$ to 4) are the ith order of dispersion terms, α is the propagation loss constant of the fiber, γ is the effective nonlinear coefficient, and

T_R denotes the Raman scattering parameter. To find the solution of the NLSE, the split-step Fourier approach is utilized, incorporating dispersion terms up to the fifth order. This method takes into account the nonlinear response $\gamma\ (w)$ that is provided by,

$$\gamma = \frac{n_2 w_0}{cA_{\text{eff}}} \tag{9.11}$$

where c is the speed of light, w_0 is the angular frequency, and $n_2 = 5.9 \times 10^{-19}$ is the nonlinear refractive index [32]. The effective area (A_{eff}) was determined using the surface integral of the mode field distribution,

$$A_{\text{eff}} = \frac{\left(\int^{|F(x,y)|^2} dxdy \right)^2}{\int^{|F(x,y)|^4} dxdy} \tag{9.12}$$

9.5.1 Supercontinuum Generation Versus Fiber Length and Dispersion

The self-phase modulation (SPM) is characterized by a nonlinear phase shift,

$$\phi_{NL}(L, T)\alpha L_{\text{eff}} \tag{9.13}$$

The Eq. (9.13) indicates that the SPM effect is directly proportional to the effective length of the fiber, denoted as L_{eff}. The simulation utilized a hyperbolic secant pulse as the input pulse, characterized by a pulse width of 100 femtoseconds (full width at half maximum), peak power of 2 kilowatts, and a repetition rate of 1 kilohertz. Figure 9.2a displays the spectrum changes of supercontinuum for the HIC Bragg fiber length. SPM primarily caused the initially observed spectral broadening. New spectral components emerged approximately 3 cm into the propagation after covering a distance of 2 cm. Spectrum intensity increased due to dispersive wave creation during propagation from 2 to 15 cm, with spectral components around 2 cm being amplified after 2 cm of transmission. Figure 9.2b displays the bandwidths of the supercontinuum spectra produced in the Bragg fiber when pumped at the zero dispersion wavelength of 1.064 μm. The fiber length is 15 cm and the power used is 2 kW. As expected, the SC spectra exhibited a rapid widening accompanied by a red-shift towards the shorter wavelength region. By pumping at the zero dispersion wavelength (ZDW), a SC spectrum with a range of 0.5–5.6 μm was obtained. An observable enhancement in the blue and red shifts of the spectral widening is observed as the propagation distance increases. The primary cause of the first spectrum broadening was predominantly due to Self-Phase Modulation (SPM). Anomalous dispersion soliton fission and Raman soliton self-frequency shift during propagation in the anomalous domain are the main factors responsible for the expansion of the spectral range at longer wavelengths. This leads to the creation of a large supercontinuum spectrum. The observed spectrum ranges exhibited minimal variation following a propagation distance of 15 cm when subjected to a ZDW of 1.064 μm. A fiber length of 15 cm resulted in the highest spectral bandwidth of the supercontinuum. The expansion of spectral broadening in the fiber was impeded by the reduction of transmission power. The dispersion and nonlinear coefficient of HIC Bragg fiber are 0.23 ps/(km•nm) and 550w^{-1} km^{-1} at a wavelength of 1.064 μm, respectively. Having a moderate length and distribution of fibers is preferred for generating supercontinuum with a broad spectral range. When the laser with pumped energy is near the zero dispersion wavelength (ZDW), the supercontinuum curve exhibits excellent flatness.

9.5.2 *Supercontinuum Generation Versus Pump Wavelength*

As demonstrated in Table 9.1, the discussion above pertains to the dispersion and nonlinear coefficient of the Bragg fiber with specific parameters: $d/\Lambda = 0.15$, $\Lambda = 1.2$ μm, dn = 0.02, $R = 5$ μm, $N = 6$. This fiber experiences an increase in dispersion value and a decrease in nonlinear coefficient as the pump wavelength is increased from 0.8 to 2 μm. The ZDW exhibited clarity at a wavelength of 1.064 μm within in the normal dispersion domain. The SC spectral evolutions exposed to these pump lasers are illustrated in Fig. 9.3. The spectral broadening of supercontinuum (SC) was investigated in a 15-cm fiber using a pump power of 2 kW and various wavelengths including 0.8, 0.98, 1.064, 1.55, and 2 μm. In the usual dispersion regime, when the fiber is pumped at 0.8 and 0.98 μm, it is necessary to provide a higher amount of energy in order to achieve a broadband supercontinuum. This is because the pump wavelength is at a short wavelength. The spectral coverage of the

Table 9.1 Values of the Bragg fiber's nonlinear and dispersion coefficients, $d/\Lambda = 0.15$, $\Lambda = 1.2$ μm, $dn = 0.02$, $R = 5$ μm, $N = 6$ [29]

Pump wavelength (nm)	Dispersion (ps/nm/km)	Nonlinear coefficient (w^{-1} km^{-1})
800	12.66	882
980	3.88	636
1064	0.23	550
1550	7.53	261
2000	114.38	157

Reprinted with permission from Publisher, Taylor & Francis Ltd., http://www.tandfonline.com: [29]

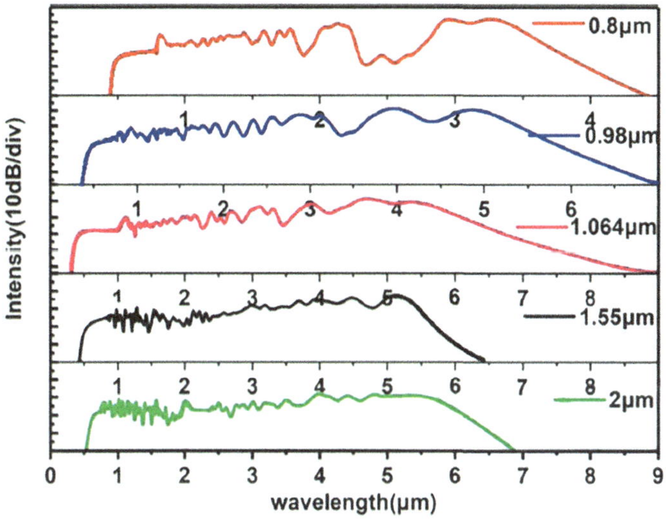

Fig. 9.3 The spectral evolutions of the SC were examined using various pump wavelengths (0.8, 0.98, 1.064, 1.55, and 2 μm) while maintaining a consistent pump power of 2 kW [29]. (Reprinted with permission from Publisher, Taylor & Francis Ltd., http://www.tandfonline.com: [29])

supercontinuum (SC) ranged from 0.49 to 3.4 μm with a pump wavelength of 0.8 μm, and from 0.5 to 4.8 μm with a pump wavelength of 0.98 μm. High power and usual expectations from the SPM interaction in the fiber are necessary for broadband SC generation. A 1.064 μm pump laser leads to an expansion of the SC spectrum from 0.5 to 5.6 μm when it is in close proximity to the ZDW. This expansion leads to a notable augmentation in the long wavelength range of the SC spectrum. Moreover, within the short wavelength range, the dispersion is rather weak, and the widening of the spectral line is highly significant. This is sufficient for traversing the Zero Dispersion Wavelength (ZDW) and facilitating the passage of energy to a broader range of wavelengths. The phenomenon of soliton dynamics subsequently causes the spectrum to exhibit an expansion towards longer wavelengths, hence giving rise to a significant asymmetric spectral broadening.

Shifting the pump wavelength to 1.55 μm in the long wavelength range resulted in the broadest supercontinuum spectrum ranging from 0.51 to 5.3 μm, with a matching dispersion value of 7.53 ps/nm/km. With an increase in pump wavelength to 2 μm, the supercontinuum spectral range expanded from 0.8 to 5.5 μm. The spectral bandwidth of the supercontinuum exhibited a decrease when the pump wavelength was increased from 1.55 to 2 μm, primarily attributed to the diminished nonlinear coefficient. The utilization of anomalous dispersion and soliton dynamics facilitated a significant expansion of the SC spectrum to encompass longer wavelengths. This was accomplished through the fission of higher-order solitons and the Raman-induced soliton self-frequency shift [33, 34].

9.6 Conclusion

This chapter briefly touches upon various nonlinear aspects of Bragg Fibers useful for photonics applications. The study of solitons in a dispersive ring that does not experience any loss and demonstrates a Kerr nonlinearity is first attempted. A brief summary of the nonlinear transmission of optical pulses having a duration of several hundred femtoseconds in solid-core Bragg fibers is then presented. The influence of pulse width, peak power, soliton order, and higher order dispersion on the development of sech pulses in the nonlinear domain was the subject of an investigation that was carried out. Theoretical models indicate that the initial broadening is mostly influenced by the effects of group velocity dispersion and self-phase modulation, whereas the creation of dispersive waves becomes evident after a specific length of the fiber. The solid-core Bragg fibers (and DDBFs) can be created using the established MCVD technology. This technique is reproducible and can be standardized, making it an economically viable platform for studying nonlinear optical effects at submicrometer wavelengths with various dispersion characteristics.

An interesting demonstration of the propagation of megawatt-class pulses in flexible all-silica fibers has been studied. It has been shown that well-designed all-glass antiresonant fibers, when pumped by fiber-based chirped pulse amplification (CPA) devices, enable the efficient creation and delivery of megawatt-class pulses over large distances. These type of Bragg fibers could enable the development of miniaturized endomicroscopes used for various biomedical applications.

Finally, a novel form of tellurite high index core Bragg fiber has been successfully produced as an innovative method for manufacturing a highly concentrated supercontinuum. High-index-core Bragg fibers possess significant potential for achieving the creation of extremely coherent mid-infrared supercontinuum. The generation of a mid-infrared supercontinuum in a high index core Bragg fiber has been numerically shown with the application of a hyperbolic secant pulse with precise parameters.

References

1. E. Yablonovitch, Photonic band-gap structures. J. Opt. Soc. Am. B Opt. Phys. **10**(2), 283–295 (1993)
2. R.F. Cregan, B.J. Mangan, J.C. Knight, T.A. Birks, P.S. Russell, P.J. Roberts, D.C. Allan, Single-mode photonic band gap guidance of light in air. Science **285**(5433), 1537–1539 (1999)
3. R.W. Boyd, *Nonlinear Optics*, 2nd edn. (Academic, San Diego, 2003)
4. S. Dasgupta, B.P. Sonali, M.R. Shenoy, Nonlinear spectral broadening in solid-core Bragg fibers. J. Lightwave Technol. **25**(9), 2475–2481 (2007)
5. D. Arbel, M. Orenstein, Self-stabilization of dense soliton trains in a passively mode-locked ring laser. IEEE J. Quantum Electron. **35**(6), 977–982 (1999)
6. Y.C. Chow, M.J. Ablowitz, The periodic cubic Schrodinger equation. Stud. Appl. Math. **65**, 113–158 (1981)

7. A. Meccozi, Long-distance soliton transmission with filtering. J. Opt. Soc. Am. B **10**, 2321–2330 (1993)
8. M. Romagnoli, M. Midrio, P. Franco, F. Fontana, Maximum soliton-train duty cycle in harmonically mode-locked fiber lasers. J. Opt. Soc. Am. B **12**, 1732–1740 (1995)
9. G.P. Agrawal, *Non-Linear Fiber Optics* (Academic, San Diego, 1989)
10. K. Thyagarajan, S. Diggavi, A. Taneja, A. Ghatak, A simple numerical technique for the analysis of cylindrically symmetric refractive index profile optical fibers. Appl. Opt. **30**(27), 3877–3879 (1991)
11. K. Morishita, Numerical analysis of pulse broadening in graded index optical fibers. IEEE Trans. Microw. Theory Tech. **MTT-29**(4), 348–352 (Apr. 1981)
12. P.K.A. Wai, C.R. Menyuk, Y.C. Lee, H.H. Chen, Nonlinear pulse propagation in the neighborhood of the zero-dispersion wavelength of monomode optical fibers. Opt. Lett. **11**(7), 464–466 (1986)
13. K. Mori, H. Takara, S. Kawanishi, M. Saruwatari, T. Morioka, Flatly broadened supercontinuum spectrum generated in a dispersion decreasing fibre with convex dispersion profile. Electron. Lett. **33**(21), 1806–1808 (1997)
14. T. Okuno, M. Onishi, M. Nishimura, Generation of ultra broadband supercontinuum by dispersion flattened and decreasing fiber. IEEE Photon. Technol. Lett. **10**(1), 72–74 (1998)
15. F. Lu, Y. Deng, W.H. Knox, Generation of broadband femtosecond visible pulses in dispersion-micromanaged holey fibers. Opt. Lett. **30**(12), 1566–1568 (2005)
16. S. Février, R. Jamier, J.-M. Blondy, S.L. Semjonov, M.E. Likhachev, M.M. Bubnov, E.M. Dianov, V.F. Khopin, M.Y. Salganskii, A.N. Guryanov, Low-loss singlemode large mode area all-silica photonic bandgap fiber. Opt. Express **14**, 562–569 (2006)
17. H. Delahaye, G. Granger, D. Gaponov, L. Lavoute, S. Aleshkina, M. Salganskii, A. Hideur, M. Likhachev, S. Février, Megawatt solitons generated above 2000 nm in Bragg fibers. Opt. Lett. **44**(11), 2713–2715 (2019)
18. J.W. Nicholson, J.M. Fini, A.D. Yablon, P.S. Westbrook, K. Feder, C. Headley, Demonstration of bend-induced nonlinearities in large-mode-area fibers. Opt. Lett. **32**, 2562–2564 (2007)
19. H. Delahaye, C.H. Hage, S.M. Bardet, I. Tiliouine, G. Granger, D. Gaponov, L. Lavoute, et al., Generation of megawatt soliton at 1680 nm in very large mode area antiresonant fiber and application to three-photon microscopy. J. Opt. **23**(11), 115504 (2021)
20. N.G. Horton, K. Wang, D. Kobat, C.G. Clark, F.W. Wise, C.B. Schaffer, C. Xu, In vivo three-photon microscopy of subcortical structures within an intact mouse brain. Nat. Photon. **7**, 205–209 (2013)
21. G. Tao, S. Shabahang, H. Ren, F. Khalilzadeh-Rezaie, R.E. Peale, Z. Yang, X. Wang, A.F. Abouraddy, Robust multimaterial tellurium-based chalcogenide glass fibers for mid-wave and long-wave infrared transmission. Opt. Lett. **39**(13), 4009–4012 (2014)
22. S.T. Sanders, Wavelength-agile fiber laser using group-velocity dispersion of pulsed super-continua and application to broadband absorption spectroscopy. Appl. Phys. B Laser Opt. **75**(6), 799–802 (2002)
23. A. Labruyère, A. Tonello, V. Couderc, G. Huss, P. Leproux, Compact Supercontinuum sources and their biomedical applications. Opt. Fiber Technol. **18**(5), 375–378 (2012)
24. A. Al-Kadry, M.E. Amraoui, Y. Messaddeq, M. Rochette, Two octaves mid-infrared Supercontinuum generation in As2Se3 microwires. Opt. Express **22**(25), 31131–31137 (2014)
25. P.F. Ambrico, A. Amodeo, P.D. Girolamo, N. Spinelli, Sensitivity analysis of differential absorption Lidar measurements in the mid-infrared region. Appl. Opt. **39**(36), 6847–6865 (2000)
26. A. Blanco-Redondo, C. Husko, D. Eades, et al., Observation of soliton compression in silicon photonic crystals. Nat. Commun. **5**, 3160 (2014)
27. T. Cheng, K. Nagasaka, T.H. Tuan, et al., Mid-infrared Supercontinuum generation spanning 2.0 to 15.1μm in a chalcogenide step-index Fiber. Opt. Lett. **41**(9), 2117–2120 (2016)

28. T.D. Engeness, M. Ibanescu, S.G. Johnson, O. Weisberg, M. Skorobogatiy, S. Jacobs, F. Yoel, Dispersion tailoring and compensation by modal interactions in omniguide fibers. Opt. Express **11**, 1175–1196 (2003)
29. J. Zheng, W. Lei, Y. Qin, H. Zou, K. Yu, W. Wei, Highly nonlinear dispersion-flattened high-index-core Bragg fibres for supercontinuum generation. J. Mod. Opt. **66**(20), 2008–2014 (2019)
30. J.H.V. Price, K. Furusawa, T.M. Monro, et al., Tunable, femtosecond pulse source operating in the range 1.06-1.33μm based on an Yb3+– doped holey Fiber amplifier. J. Opt. Soc. Am. B **19**(6), 1286–1294 (2002)
31. J.M. Dudley, G. Genty, S. Coen, Supercontinuum generation in photonic crystal fiber. Rev. Mod. Phys. **78**, 1135–1184 (2006)
32. M. Liao, C. Chaudhari, G. Qin, et al., Fabrication and characterization of a chalcogenide-tellurite composite microstructure fiber with high nonlinearity. Opt. Express **17**(24), 21608–21614 (2009)
33. T. Cheng, K. Nagasaka, T.H. Tuan, et al., Mid-infrared Supercontinuum generation spanning 2.0 to 15.1μm in a chalcogenide step-index fiber. Opt. Lett. **41**(9), 2117–2120 (2016)
34. P. Yeh, A. Yariv, E. Marom, Theory of Bragg fiber. Opt. Soc. Am. **68**, 1196–1201 (1978)

SPRINGER NATURE

GPSR Compliance

The European Union's (EU) General Product Safety Regulation (GPSR) is a set of rules that requires consumer products to be safe and our obligations to ensure this.

If you have any concerns about our products, you can contact us on ProductSafety@springernature.com

In case Publisher is established outside the EU, the EU authorized representative is:

Springer Nature Customer Service Center GmbH
Europaplatz 3
69115 Heidelberg, Germany

The manufacturer's authorised representative in the EU is Springer Nature Customer Service Centre GmbH, Europaplatz 3, 69115 Heidelberg, Germany. If you have any concerns regarding our products, please contact ProductSafety@springernature.com

Printed and bound by CPI Group (UK) Ltd, Croydon, CR0 4YY
30/09/2026
02225239-0003